GREAT CALCULATIONS

ALSO BY COLIN PASK

Magnificent Principia

Math for the Frightened

GREAT CALCULATIONS

A Surprising Look Behind
50 Scientific Inquiries

COLIN PASK

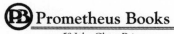

Prometheus Books

59 John Glenn Drive
Amherst, New York 14228

Published 2015 by Prometheus Books

Cover image (top) © Bigstock
Cover image (bottom) © Media Bakery
Cover design by Grace M. Conti-Zilsberger

Inquiries should be addressed to
Prometheus Books
59 John Glenn Drive
Amherst, New York 14228
VOICE: 716–691–0133
FAX: 716–691–0137
WWW.PROMETHEUSBOOKS.COM

19 18 17 16 15   5 4 3 2 1

Library of Congress Cataloging-in-Publication Data

Pask, Colin, 1943-
    Great calculations : a surprising look behind 50 scientific inquiries / by Colin Pask.
        pages cm
    Includes bibliographical references and index.
    ISBN 978-1-63388-028-3 (pbk.) — ISBN 978-1-63388-029-0 (e-book)
    1. Mathematics—Miscellanea. I. Title.

QA99.P385 2015
510—dc23

                                                                                    2015001967

Printed in the United States of America

To Melanie and Daniel

CONTENTS

ACKNOWLEDGMENTS

My interest in the calculations discussed in this book goes back fifty years, and over that time a great many colleagues have helped and inspired me. My doctoral-thesis supervisor, John M. Blatt, was a pioneer in the use of computers for tackling physical problems, and his enthusiasm and wisdom set me off on my career. Barry Ninham met my wife and me when our boat arrived in Australia fifty years ago, and he has been a friend and supporter ever since. As a highly skilled applied mathematician, he will recognize his influence on me with the appearance of Bessel functions in this book. The wonderful Annabelle Boag has produced the magnificent figures that grace this book; I would be lost without her cheerful support and highly professional assistance. Peter McIntyre is a meticulous reader of manuscripts, and I thank him for much help and advice. Connie Wintergem has supported me in much of the research for this book. My beautiful wife Johanna has shared life with me for fifty years, and this book owes more than I can say to her love and support.

PREFACE

S cience combines observation, experiment, and theory. The observational and experimental sides are probably better known and more readily appreciated. The spectacular observations of astronomers, for instance, are often in the news. Many people have read about or have seen television reports on the 2012 experiment at the Large Hadron Collider in which particles were smashed together to reveal the existence of the Higgs boson. But it is likely that few of these people understood why the experiment was carried out and what the result means for the future of science. It is the interplay between experiment and theory that gives sciences its coherence, integrity, and power. This book aims to improve the understanding of the role played by theory in science. It complements the many books attempting to identify science's most influential experiments by describing a selection of its great calculations.

The calculations described in this book are mostly relatively simple (at least in principle, if not always in practice) but still of major significance. I believe they are accessible to the ordinary reader; occasionally the more technical details are given in separate sections for those wishing to see them. In any case, it is usually the context which is of major importance: Why was the calculation made? What was the result? What was the effect of the calculation on the progress of science? Did the result have broader implications for society and the future of mankind? Answers to questions such as these reveal how calculations are both important and interesting— and also sometimes controversial, as the predictions for climate change illustrate.

The calculations are grouped under broad subject areas such as the solar system or the building blocks of matter. In each group they are presented in chronological order, and in this way readers may learn something of the history of science (although the different parts of the book may be read in any order and dipped into as desired). Similarly, I illustrate the ways in which the styles and techniques used in calculations have evolved, something I review in the final chapter and which may be of particular interest to those readers who have grown up entirely in the computer age. We shall also meet some of the great characters involved, from Archimedes to Einstein, and discover the intriguing stories surrounding certain calculations.

By reading about the various calculations I hope that you will come to better appreciate the vital link between experiment and theory.

It will be obvious that the choice of calculations to be described reflects my personal interests, background, and views of science. It was not easy to keep to such a restricted number of examples (and some extras have briefly sneaked in as "honorable mentions"). I expect some readers may be upset not to find their personal favorites on my list. Even more provocative may be my final choice of the ten calculations most worthy of the label "great." However, that is part of the fun. I stand ready to be informed and educated about the superior qualities of other worthy candidates I have excluded.

FIFTY QUESTIONS FOR THE CALCULATIONS TO ANSWER

1. How did a parson warn us about the problems of overpopulation?
2. How mathematically advanced were the ancient Babylonians?
3. What could have been another "eureka moment" for Archimedes?
4. What did Fibonacci do apart from showing how rabbit populations get out of control?
5. How do you double the life span of an astronomer?
6. Can you add up an infinite number of things?
7. How did Gauss find a mysterious formula for prime numbers?
8. How big is the earth?
9. How old is the earth?
10. What is inside the earth?
11. How did Galileo mathematically shoot a cannon?
12. Which nineteenth-century calculation used an analogue computer that was then used in planning the allies' D-Day landings in France?
13. Which masterpiece dominated astronomy for a thousand years?
14. How did Kepler battle with Mars?
15. Why is the link between a falling apple and a falling moon so important?
16. How do you weigh the planets?
17. How did Halley predict the return of his comet?
18. How do you find a new planet without using a telescope?
19. How far away is the sun?
20. Which calculation led Einstein to say that "for a few days, I was beside myself with joyous excitement"?
21. Why is the sky dark at night?
22. How old is the universe?

23. How does carbon-based life on Earth depend on a quirk of nuclear physics?
24. How do we know there is lots of "dark matter" in the universe if we can't see it?
25. How did a parson's dark stars turn into black holes?
26. How did a little calculation of pumping by the heart overturn over a thousand years of medical practice?
27. What links a famous comet and the cost of an annuity?
28. How did a pure mathematician who claimed to have never done anything "useful" give his name to one of the first major results in genetics?
29. Why is the CT scan unfairly named?
30. What links a mouse to an elephant?
31. How can Jupiter's moons tell you the speed of light?
32. How did Newton upset the poet John Keats?
33. What limits the details you can see?
34. How did Maxwell see the light?
35. What is light?
36. Could Einstein's light bending prove his "Dear Lord" wrong?
37. How do you know atoms really exist?
38. How did a Swiss schoolteacher find patterns in light?
39. How did a French prince explain the schoolmaster's patterns?
40. How can you measure the mass of a neutron?
41. What is the most accurate calculation of all?
42. How did a threat to the conservation of energy law lead to a John Updike poem?
43. Can you stay younger by going faster?
44. Why should you believe in quarks?
45. Why does the sun shine?
46. How do you plan to make the biggest bomb of all?
47. Who discovered a mathematical poem for violin strings?
48. How were those amazing mathematical tables made?
49. How do you ask a computer about a string of oscillators?
50. How easy is it to find chaos?

Chapter 1

INTRODUCTION

in which I review the complementary roles of calculation and experiment in science; outline some general aspects of calculations and see them in operation in a first example; and explain the structure of the chapters to follow.

According to Einstein,

> science is the attempt to make the chaotic diversity of our sense experience correspond to a logically uniform system of thought. In this system single experiences must be correlated with the theoretic structure in such a way that the resulting coordination is unique and convincing.[1]

In science, we investigate *sense experiences* in a systematic way. This may involve planned suites of observations in astronomy or in zoology, or the arrangement of particular physical entities and conditions in experiments. The importance of the use of both theory and experiment was neatly summarized long ago by philosopher Immanuel Kant (1724–1804):

> Experience [experiment] without theory is blind, but theory without experience is mere intellectual play.[2]

(By the way, we must not be too disparaging about intellectual play because it might be said to cover pure mathematics, where the formalism itself rather than its applications is of interest; we shall see some famous calculations within this category, too, before we move on to applying mathematics in science.)

The development of this viewpoint by people like Sir Francis Bacon, Galileo, John Locke, Sir Isaac Newton, and Kant was central to the scientific revolution underway by the seventeenth century. Nevertheless, its recognition was still in question for many years as is apparent from this excerpt from an 1861 letter from Charles Darwin to Henry Fawcett:

> How profoundly ignorant B. must be of the very soul of observation! About thirty years ago there was much talk that geologists ought only to observe and not theorize; and I well remember someone saying that at this rate a man might as well go into a gravel-pit and count the pebbles and describe the colors. How odd it is that anyone should not see that all observation must be for or against some view if it is to be of any service![3]

Today, it is generally accepted that we need theory to discover order in the mass of data revealed by experiments and also when planning those experiments (in Einstein's words, *theory tells us what to measure*[4]). However, while there are many books about the outstanding, or most important, experiments (see those by Crease, Johnson, and Shamos for examples), very few deal with the actual calculations behind such experiments. There are books about the equations of physics (see Crease, for example), but equations are the tools of theory—much like the telescope, cyclotron, and oscilloscope are tools in experimental physics. Here I am concerned with the most influential results of using these theoretical tools.

One reason for the theory-experiment imbalance is that to treat the contribution of theory in the physical sciences often means coming to terms with a mathematical formalism. Mathematics has proved itself to be the tool required by scientists; for Galileo there is a *mathematical language* for the universe.[5] Mathematics allows us to use and interpret observations to "go where we cannot go"—to explore the outer reaches of the universe, the interior of our planet, and the subatomic world—and even reveals what is inside our own heads.

Kant's idea that *theory without experience is mere intellectual play* is at the heart of a debate about the value of string theory in present-day fundamental theoretical physics.[6] There is a danger that the search for ever more fundamental theories loses contact with experiment and the actual physical universe as we know it; for science to prosper there must be calculations and a direct comparison with observations.

I contend that many calculations in science may be appreciated without a deep knowledge of the mathematics involved, just as the output from an experiment may be appreciated without understanding the intricacies of apparatus design and manipulation. This book gives a discussion of calculations to go along with those books that describe the experimental side.

1.1 ABOUT CALCULATIONS

To put the mathematical aspects into perspective, I suggest that four questions should guide the discussions:

- Why was the calculation made?
- What was calculated?
- What was the result of the calculation?
- What impact did the calculation have?

The details of exactly how the calculation was made may be of secondary importance, although in some cases the approach and techniques used are so innovative or revolutionary that they do merit discussion. Mostly, if I do give technical details, I will put them in separate sections for those who wish to see them, and always in the bibliography I give references for readers who wish to study the original papers reporting the calculations.

We shall see how calculations help to turn data into information and thus fit them into the patterns of science. Sometimes single numbers are involved. Sometimes whole data sets, perhaps very extensive in nature, are to be treated. The calculation may result in a formula or parameters to be used in a descriptive formula. This often reduces experimental results to a form that requires explanation and lends itself to interpretation; one calculation may inspire another. Sometimes the formula is already suggested by theory, and then the calculation will be validating or discrediting that theory.

Remembering Einstein's *theory tells us what to measure*, we shall also see examples where the calculation produces a prediction for experimental verification. As mentioned above, this is the key step in real science: put the theory in a form that can be tested by experiment. When Einstein's general theory of relativity was published, it contained one calculation explaining a mysterious anomaly in planetary motion measurements and another predicting an optical effect. For many people, it was the prediction that was so impressive because there is always a lingering feeling that a theory may be manipulated to produce calculations of already-known effects. (Of course there is then the related question of whether data may be selected or doctored to fit a prediction.[7])

An example will help to clarify some of these matters.

1.2 EXAMPLE: MALTHUS GIVES US CALCULATION 1

I begin with a very simple calculation: generate two sequences of numbers, each starting with the number 1. In one case, go from one term to the next by doubling the term; in the other, just add 1 to the term. We obtain

1 2 4 8 16 32 64 . . .
1 2 3 4 5 6 7 . . .

The first one is an example of a geometric sequence (each term is just a multiple of the previous one and here the multiplying factor is 2). The second one is an example of an arithmetical sequence (each term is a sum of the previous one and a constant factor, and here that factor is 1). Not a very profound result, but we shall see more exciting numerical examples in the next chapter.

The interest in these two sequences increases dramatically when we note their use in Parson Thomas Malthus's *Essay on the Principle of Population as It Affects the Future Improvement of Society* (1798). Malthus begins with two *postulata*:

First, that food is necessary to the existence of man.

Second, that the passion between the sexes is necessary, and will remain nearly in its present state.[8]

He goes on to deduce that *population, when unchecked, increases in a geometrical ratio. Subsistence increases only in an arithmetical ratio.* He then points out: *a slight acquaintance with numbers will shew the immensity of the first power in comparison of the second.* The numbers he uses as examples are just those given above. To make things clear, I define time periods (with zero indicating the initial situation), and the population and the available resources in those periods. Those resources can provide for a certain population, and I calculate that supportable population as three times the resources value. You might think of something like a number of years (Malthus assumed 25 years), millions of people, and thousands of acres of farming land. Malthus's results then appear as follows:

time period	0	1	2	3	4	5	6 . . .
population	1	2	4	8	16	32	64 . . .
resources	1	2	3	4	5	6	7 . . .
supportable population	3	6	9	12	15	18	21 . . .

Clearly life is good initially; the resources are more than adequate for supporting the population. But by period 4 the population is barely making ends meet. Life becomes very tough in period 5 (perhaps surviving by using stored food excesses from previous periods), and by period 6 there is complete disaster with the actual population more than three times the supportable one. In Malthus's own words:

This implies a strong and constantly operating check on population from the difficulty of subsistence. This difficulty must fall somewhere; and must necessarily be severely felt by a large portion of mankind.

I label this **calculation 1, Malthus on population growth**. By means of his simple calculation Malthus has identified a problem (the failure to match resources to increasing population) that has troubled mankind in almost all civilizations and continues to cause misery in much of the world today. The value of the calculation is in dramatically highlighting the problem.

1.2.1 Observations

Notice that we have now seen a common pathway for calculations with the following steps:

1. State the problem.
2. Identify the defining elements.
3. Translate the problem into a mathematical form.
4. Determine how to carry out the calculation.
5. Obtain the required result.
6. Analyze the result to see what it is telling us.

Malthus's example also illustrates another property of calculations: sometimes the objective is to identify a trend or type of behavior; in this case the way population growth outstrips the availability of resources needed to support it. The actual detailed numbers (population doubling in 25 years for Malthus) are not of particular interest; it is the pattern they reveal that is of importance.

Of course, some people will always want a more realistic assumption, as Charles Babbage amusingly demonstrated.[9] After reading the couplet "Every minute dies a man, Every minute one is born," in Alfred Tennyson's *Vision of Sin*, Babbage wrote to Tennyson suggesting that it be changed to "Every minute dies a man, And one and a sixteenth is born." But even that was not enough as he continued "I may add that the exact figures are 1.167, but something must, of course, be conceded to the laws of metre." At least Babbage wished to build in the trend of an increasing population, rather than the static one as Tennyson would have it.

Malthus's calculation is a good example of one that produces a set of numbers. However, sometimes a calculation may result in a formula. For example, if we denote the population in time period n by p_n, then Malthus tells us that $p_n = 2^n$. The formula may be one to be tested against experimental results, and its accuracy may be a test of the validity of the underlying theory. On other occasions, the formula may be deduced by finding the appropriate mathematical expression to describe data from natural observations or from carefully designed experiments.

It is a feature of many good calculations that they lead us to ask questions and to build in other characteristics of the problem. In chapter 12 we will see how Malthus's work leads into whole new areas of both population modeling and mathematics.

It may be that an original calculation fails to take account of some essential aspects of the situation under investigation, perhaps something unknown at the time when it was made. This would invalidate any conclusions to be drawn from the calculation. We will see some famous examples in later chapters.

The results of a calculation might also lead us to different but related questions; in Malthus's case: How many people can the earth support? In fact, this question had been considered by Antonie van Leeuwenhoek (1632–1723), a man better known as a pioneer in microscopy. (Details are given by Cohen, who explains how Leeuwenhoek came up with a maximum population of 13,385,000,000.)

Finally, we should note that Darwin was aware of the conclusions to be drawn from Malthus's work when he considered how the essential competition for scarce or limited resources might be incorporated into his theory of evolution. Calculations may clearly have an influence far beyond their immediate context.

The actual calculations made by Malthus may have been extremely simple, but they have led to some vitally important developments in science and in the affairs of mankind.

1.3 OUTLINE FOR THE BOOK

I have grouped my chosen calculations into broad subject areas, and within each group, I use chronological ordering so we can see how calculations sometimes mark a major milestone in the history of science. Sometimes

I will refer to a group of calculations. Emphasis is placed on the physical sciences, since it is there that mathematical descriptions and calculations have long been accepted as part of the subject, but examples from the biological sciences are also given. However, first, in chapters 2 and 3, I present some examples of calculations that are of purely mathematical interest, although in almost every case there is also a link to the applied mathematical world.

Chapter 4 shows how calculations have helped us to get to know our home, the earth. Calculations are becoming ever more important (think weather forecasting and climate change), and I will touch on some issues involved in a discussion section.

Chapters 5, 6, and 7 examine the part played by calculations when we consider our earth as a planet in a larger framework, first in the solar system and then in the universe as a whole. It is here that calculations become of necessity and exceptional importance.

Chapter 8 moves from the physical to the biological (having already seen Malthus's calculations earlier in this chapter). The physical world seems to demand the use of mathematics when we study it, but the triumphs of such an approach are fewer—but still vitally important—in the biological domain, and I will speculate a little about that.

The next three chapters deal with the basic and most fundamental parts of our physical world: light and the building blocks of matter. Discussion of the latter subject is divided into two chapters since nuclear physics provides some of the most spectacular (and perhaps disturbing) examples of the impact of calculations.

Chapter 12 examines how calculations have allowed us to discover the nature of motion in its various forms and to see how general mathematical descriptions were developed. It is here that we shall see the impact of modern computers as a vital ingredient in the business of calculation.

Finally comes a chapter of summary and evaluation. Can we say how, and to what extent, calculations have molded and supported the progress of science? Even more contentiously, can we pick out those calculations that are of supreme importance, those worthy of the label "great calculations"? It is an evaluation of their history, context, and impact that contributes to an assessment of a particular calculation, and here opinions are bound to differ. Please note that the title of this book is *Great Calculations* and not *THE Great Calculations*, or some other title suggesting that there is a definitive list. This is surely a personal matter, and many readers may have a quite different list. That is part of the fun, and I ask you not to be too irate if your personal favorite is missing.

References for all chapters are given in the notes section and also in the bibliography, which contains details of all mentioned books and papers. I have given examples of suitable references so that the interested reader may further explore each calculation.

If you cannot wait for the story to unfold, but want to see my list of important calculations right now, please go to the beginning of chapter 13!

Chapter 2

ANCIENT MATHEMATICS

*in which we look at two outstanding examples of
calculations made a long time ago.*

Studies in cognitive science have revealed that humans (and some
other animals) have certain innate mathematical abilities. At the
simplest level, we readily distinguish the difference between
groups of small numbers of objects (the power of numerosity) and recognize primitive geometric shapes and properties. This is linked to the
natural world and our need to understand it in order to survive and prosper.
Further development leads to arithmetic and simple ideas of measurement
and design. Humans have also developed mathematical concepts and procedures that go beyond that link to the natural world. We speak of pure
mathematics, with its emphasis on the mathematical formalism itself, as
opposed to applied mathematics, where that formalism is exploited in
order to describe the physical world. However, there is a continual interplay between pure and applied mathematics as problems and techniques
are suggested and improved. Many people find it quite mysterious that
abstract mathematical work later turns out to be useful in science.

This chapter and the next describe some calculations that are basically
in the realm of pure mathematics, but that back-and-forth dialog between
such calculations and their applications is ever present and will be nicely
illustrated. I have chosen two examples from the work of ancient mathematicians for discussion in this chapter.

2.1 HISTORY BEGINS: TO MESOPOTAMIA FOR CALCULATION 2

In the period from around 3000 BCE, civilizations prospered in Mesopotamia, the region between the Tigris and Euphrates Rivers that we know today as Iraq. The region is sometimes called the cradle of civilization. People gathered in large towns, and city-states evolved. Major building and construction projects were undertaken, and farming using irrigation systems allowed large populations to be fed. There was development of ideas and practices in many areas: administration, planning, legal, military, trading, and religion. Naturally there was also the development of methods and materials for keeping records of all these things.

In one of the most significant of all steps in civilization, the early Sumerians (and later the Babylonians, Assyrians, and Hittites) developed cuneiform writing. A reed stylus was pressed into clay tablets to produce records that, after some process of drying or baking, became permanent. (The book by Robinson gives an introduction to writing and its importance.) Thousands upon thousands of ancient Mesopotamian clay tablets survive today. So it is that prehistory first moves over to history, as writing allows knowledge of life in those times to be obtained directly from the records left by the people themselves.

Many of the tablets are concerned with matters of everyday life such as administration, commerce, and scribal teaching and practice work. But tablets also give us what is surely the first great work of literature, *The Epic of Gilgamesh*. Even today the story of the tyrant king is fascinating to read. In Andrew George's translation of the work, you can see many examples of cuneiform tablets and an appendix explaining the translation process. Legal codes were developed, culminating in the 282 laws set down by Hammurabi around 1750 BCE. The Code of Hammurabi may be read in surviving cuneiform inscriptions, and we can appreciate how justice was meted out fairly, if somewhat harshly.

According to Jerry Brotton in his *A History of the World in Twelve Maps*, the very first map of the world may be viewed on a tablet from Sippar in southern Iraq. The tablet is now held in the British Museum (see bibliography). We can see how geometrical figures—circles, triangles, and rectangles—are carefully drawn. There is a hole presumably left by a pair of compasses used to draw the large circles. The cuneiform text explains the map; for example, beyond the outer circle is a "salt sea" representing the ocean encircling the inhabitable world.

Naturally the worlds of commerce, engineering, and administration, as well as the extensive investigations of astronomy and astrology initiated in Mesopotamia, led to the invention of mathematical ideas and their use in problem solving. (Otto Neugebauer is the classic reference.) Many thousands of tablets tell us that these people were skilled in mathematics, and they produced tables of all kinds. For numerical work, they counted to ten and then used multiples of ten as we do today. However, when they reached sixty they changed over to use 60 as the base for their numbers. As an example of the sexagesimal system, where we write:

$$1565 = 1 \times 10^3 + 5 \times 10^2 + 6 \times 10 + 5,$$

those ancient mathematicians wrote:

$$\underline{\underline{425}} = \underline{\underline{4}} \times 60^2 + \underline{\underline{2}} \times 60 + \underline{\underline{5}}.$$

I have used the double underline to emphasize that they used cuneiform symbols, not the 4, 2, and 5 that we use today. Similarly their fractions were considered in terms of 60 rather than our decimal ten. For example, while we might use things like 7/100, they worked with expressions like 7/60 or 11/360. Today we still retain this ancient use of base 60 in our measurement of angles in terms of degrees, minutes, and seconds.

Tablets reveal how ancient peoples frequently used mathematics in practical applications and in all sorts of problems and puzzles, much as we still do today. It is possible to see how mathematics (as well as other disciplines) was taught, and how many problem sets, as well as worked examples and useful tables, were recorded. Enough tablets were discovered in one ruined building in Nippur to suggest it was a school. (The interested reader is referred to Eleanor Robson's highly informative and entertaining article, "Mathematics Education in an Old Babylonian Scribal School.")

There is one clay tablet that has achieved great fame, and I have chosen the mathematics it records as my second noteworthy calculation.

2.1.1 Plimpton 322

Around 1923, George Plimpton purchased the clay tablet shown in figure 2.1. It is now labeled as "Plimpton 322" in the collection at Columbia University. Physically, it is approximately 12.5 cm by 8.8 cm, and the left-hand edge indicates that a portion has broken off and been lost. There are headings and fifteen rows of numbers set out in four columns. The reverse side is blank, but may have been reserved for an extension of the given table (see below). The tablet was probably written around 1800 BCE in the ancient city of Larsa in present-day Iraq. (There is a large literature based on Plimpton 322, with early work by Neugebauer, later work by Eleanor Robson, and the recent major review by Britton, Proust, and Shnider, and references therein, providing a wealth of detail and discussion.)

Figure 2.1. The clay tablet known as Plimpton 322. *Courtesy of the Plimpton Collection, Rare Book and Manuscript Library, Columbia University.*

The table on Plimpton 322 concerns Pythagorean triples. (I shall continue to use this term even though the work in question predates Pythagoras by a very long time.) Recall that three integers (*a, b, c*) form a Pythagorean triple if

$$a^2 + b^2 = c^2. \tag{2.1}$$

Therefore, choosing the smallest possible numbers gives (3, 4, 5) since $3^2 + 4^2 = 25 = 5^2$. Are there any more Pythagorean triples? After a little experimentation you might come up with (5, 12, 13) since $5^2 + 12^2 = 169 = 13^2$. But how far would simple experimentation take you beyond that? Would you be likely to find the triple (4601, 4800, 6649), for example?

Obviously the name comes from the fact that *a, b,* and *c* could be the lengths of the sides of a right-angled triangle with hypotenuse of length *c*. Equally well, *a* and *b* could be the side lengths of a rectangle with diagonal of length *c*, and it appears that the ancient Mesopotamian mathematicians used that interpretation. Table 2.1 sets out the relevant Plimpton 322 examples.

$(c/b)^2$	*a*	*b*	*c*	*n*
1.9834	119	120	169	1
1.9492	3367	3456	4825	2
1.9188	4601	4800	6649	3
1.8862	12709	13500	18541	4
1.8150	65	72	97	5
1.7852	319	360	481	6
1.7200	2291	2700	3541	7
1.6927	799	960	1249	8
1.6427	481	600	769	9
1.5861	4961	6480	8161	10
1.5625	45	60	75	11
1.4894	1679	2400	2929	12
1.4500	161	240	289	13
1.4301	1771	2700	3229	14
1.3872	56	90	106	15

Table 2.1. Numbers related to the Plimpton 322 tablet expressed in decimal form.

The four columns on the Plimpton 322 tablet give numbers corresponding to those shown in table 2.1 but with a left out. The tablet numbers, written in cuneiform and using base 60, have been converted here to decimal form. Six errors (two of which appear to be simple copying errors) have been corrected, and obvious values for broken-off parts of the tablet have been inserted.

We see that the ancient mathematicians in Mesopotamia produced a remarkable table of Pythagorean triples. For me it represents an outstanding pioneering mathematical achievement, and without hesitation I have taken it as **calculation 2, Mesopotamian Pythagorean triples** in my list of significant calculations.

2.1.2 Discussion

A brief glance at the examples in table 2.1 tells us that this was not just some haphazard search for Pythagorean triples. Even with an electronic calculator, the reader will find it is a tedious business finding these sets of numbers, and a search starting with the smallest numbers in (3, 4, 5) will not be very helpful. Some of the numbers in table 2.1 are quite large: 18541 in row 4, for example.

Just looking at the triples on the tablet might suggest a jumble of results, maybe just stumbled upon. But how likely is it that anyone would just happen across

$$2291^2 + 2700^2 = 3541^2?$$

The numbers in column one suggest a different story; they steadily decrease and tell us that this table is not a random set of examples but instead represents the results of a systematic procedure. Some people note that, if a geometric interpretation is considered, the figures in column one give the square of the cosecant of the angle in the appropriate right-angle triangle. (The cosecant of an angle in a right-angled triangle is found by dividing the length of the hypotenuse by the length of the side opposite that angle.) However, this interpretation, while certainly valid, does not correspond to any such things in Mesopotamian mathematics. I return to this construction problem in the next section.

There has also been great debate about why Plimpton 322 was produced. It could have been as a table of results for use in solving problems of various kinds, including those based on the geometry of rectangles. It could be a table of examples for teachers to use. The interested reader will find discussions noted in the references.

2.1.3 Constructing the Table

Now, some details for those who wish to consider how the table on Plimpton 322 was constructed. There is an extensive literature on this topic, but I will present the well supported argument propounded by Britton, Proust, and Shnider. There are two points to discuss: How can Pythagorean triples be calculated? And what input should be used to produce a systematic list using that method?

If we take a number p and from it find the two numbers x and y according to

$$x = \frac{1}{2}\left(p + \frac{1}{p}\right) \quad \text{and} \quad y = \frac{1}{2}\left(p - \frac{1}{p}\right), \tag{2.2}$$

a little algebra shows that

$$y^2 + 1 = x^2. \tag{2.3}$$

Thus we have found the triple $(y, 1, x)$. For example, choosing $p = 2$ gives $(3/4, 1, 5/4)$. This is not a true Pythagorean triple since integers are required. However, multiplying all numbers in a triple gives another triple (simply scaled, we might say), and multiplying here by 4 gives us the $(3, 4, 5)$ Pythagorean triple.

The above procedure uses a number p and its inverse $1/p$. The Mesopotamian mathematicians made extensive use of inverses and produced tables of them for use in division (dividing by p is equivalent to multiplying by $1/p$) and other calculations. (See the article by Eleanor Robson.) They also had methods for the scaling process. Thus using the idea expressed in equations (2.2) and (2.3) seems like a natural approach for them to use when seeking triples.

It remains to explain the choice of values for p. Britton, Proust, and Shnider explore various published possibilities and conclude that the table is based on p values given by the regular numbers obtained by dividing the number r by the number s, r/s, with the condition that

$$1 < r/s < 1 + \sqrt{2} \qquad \text{and} \qquad 1 \le s < 1,0 \ \text{(or 60 in base 10)}.$$

Regular numbers are those having an inverse that is a sexagesimal fraction with a finite number of places, something these mathematicians would have readily tabulated (see Neugebauer's text for further details). Working up through these choices for p produces the table set out on Plimpton 322. It also shows how the detected errors would naturally arise.

If the above procedure is continued through all suitable r and s values (those values satisfying the above conditions), it will produce a total of 38 sets of triples. Britton, Proust, and Shnider give a picture of the reverse side of Plimpton 322 showing vertical lines and space (they argue) for the 23 triples which could not be fitted onto the front side of the tablet. They also suggest that the part broken off from the left edge of the tablet might contain the values of x and y.

We will never know for certain whether the above construction was actually the one used for Plimpton 322, but we can be sure that it was a systematic and beautiful piece of mathematics showing an appreciation for numbers and their properties. Mathematical thinking had begun in ancient Mesopotamia and has continued ever since. There is no historical speculation involved; the evidence is clearly written down for all to see.

2.1.4 Comments

One other related calculation is worth mentioning. It is natural to ask about the right-angled triangle with side lengths 1 which leads to the triple (1, 1, $\sqrt{2}$). Tablets similar to Plimpton 322 show that this problem was considered and that the mathematicians involved knew how to calculate $\sqrt{2}$. According to Neugebauer, they found for $\sqrt{2}$ the sexagesimal value 1,24,51,10, which in our decimal system gives 1.414213 (as against 1.414214 correct to 7 figures).

These calculations suggest the use of the "false position" or iterative methods, which allow a guess for the answer to a problem to be refined. Such methods were widespread in early mathematics, and they remain part of numerical analysis to this very day. (See chapter 3 in the book by Chabert.)

It would be wrong to see Plimpton 322 as an instigator of what today we call number theory, but it does suggest questions we might expect to find in later work. For example, we can ask: Does the above construction method generate all possible Pythagorean triples? The answer is no. The complete formalism was probably not known until the time of Euclid. Coming right up to the present, we can now answer another question suggested by the study of triples of numbers in particular relationships: What happens if we use cubes or other powers instead of squares? So we ask about $a^3 + b^3 = c^3$. In fact, there are no solutions (suitable integer values for a, b, and c) for the case of cubes or any higher powers; that result is embodied in the famous Fermat's last theorem.

Finally, we should note that the ancient Egyptian civilization was developing alongside that of Mesopotamia. There too a start was made on a range of mathematical investigations. Many examples can be seen on the famous Rhind papyrus (see the books by Calinger and Gillings), but in my opinion none of them could replace Plimpton 322 in the list of highly significant calculations. Interestingly, Richard Gillings presents evidence suggesting that the ancient Egyptian mathematicians had no knowledge of Pythagoras's theorem.

2.2 ARCHIMEDES GIVES US CALCULATION 3

The circle is the simplest of curves, and it is also the most symmetric, beautiful, mysterious, and captivating. As an example of the power of these properties, in chapter 5 we shall see how some of them actually set back science for over a thousand years. Evidence of the interest in circles can be found in ancient civilizations, and the references given for the previous section show that to be true for the Mesopotamian and Egyptian mathematicians. Moving on to around 300 BCE, we find Euclid in his *Elements* specifying only one curve (beyond the straight line) in his basic definitions:

> 15. A circle is a plane figure contained by one line such that all the straight lines falling upon it from one point among those lying within the figure are equal to one another.
> 16. And the point is called the center of the circle.
> 18. A diameter of the circle is any straight line drawn through the center and terminated in both directions by the circumference of the circle, and such a straight line also bisects the circle.[1]

We say that the distance from the center to any point on the circle is the radius r, and then the diameter has length $d = 2r$. We also define C as the length of the circumference and A as the area enclosed by the circle. (Incidentally, if we seek the closed curve of a given length enclosing the maximum area, the answer is a circle with C equal to the given length—just one of those wonderful properties of the circle.)

The natural question is now: How are a circle's circumference C and area A related to its basic size parameter r or d? Ancient mathematicians had empirical methods for estimating the area of a circle (see Gillings for Egyptian examples), but there is no evidence of a true mathematical relationship.

In the *Elements*, Euclid writes:

Book XII. Proposition 2. Circles are to one another as the squares on the diameters.

In the *Elements*, area is dealt with by comparing figures and showing when their areas are equal; there is no algebraic or numerical approach as we use today when saying, for example, a rectangle with side lengths s

and p has area s times p. Euclid obtained his circles result by considering polygons inscribed in the circles and letting the number of sides increase indefinitely in an application of the method of exhaustion.

Today we say Euclid showed that the area of a circle is proportional to the square of its diameter, or $A \propto d^2$. (Readers will probably want to say that $A = \pi r^2$ and $C = \pi d = 2\pi r$, but that is jumping ahead in the story!)

2.2.1 Enter Archimedes

Archimedes was an intellectual giant and certainly the greatest mathematician of ancient times. He was born in Syracuse in Sicily around 287 BCE. His extensive writings cover an enormous range of mathematical and scientific topics, but here we are concerned with his "Measurement of the Circle." The original no longer exists, but rough versions (which were perhaps part of a larger document) do survive. "Measurement of the Circle" contains just three propositions, and they brilliantly answer our question about the relationship between r (or d) and C and A. All three are linked together in

> Proposition 1. The area of any circle is equal to a right-angled triangle in which one of the sides about the right-angle is equal to the radius, and the base [the other side about the right-angle] is equal to the circumference.[2]

Notice that Archimedes works in the style of Euclid and expresses the area of a circle in terms of the area of another figure, this time a right-angled triangle. Today we express *Proposition 1* as

$$A = \tfrac{1}{2} rC. \qquad (2.4)$$

If we can relate the circumference to the radius, then the whole problem is solved, since then we can use the above result to give A just in terms of the radius. This is exactly what Archimedes achieves, but for some reason the propositions are given in the reverse order.

2.2.2 The Great Step

Archimedes gives

> Proposition 3. The circumference of any circle is three times the diameter and exceeds it by less than one-seventh of the diameter and by more than ten-seventyoneths.[3]

We write this as

$$\left(3 + \frac{10}{71}\right) \textit{diameter} < \textit{circumference} < \left(3 + \frac{1}{7}\right) \textit{diameter},$$

or symbolically,

$$\left(3 + \frac{10}{71}\right) d < C < \left(3 + \frac{1}{7}\right) d. \tag{2.5}$$

Many readers will not be able to resist saying that $C = \pi d$, so, in effect, Archimedes is telling us that

$$\left(3 + \frac{10}{71}\right) < \pi < \left(3 + \frac{1}{7}\right) \qquad \text{or} \qquad \frac{223}{71} < \pi < \frac{22}{7}. \tag{2.6}$$

However you wish to view this result, it is a startling and brilliant one. Archimedes has told us that the circumference of a circle is a certain number times its diameter, and he has given bounds on that number. The upper bound of 22/7 has always been commonly used as a good, simple approximation for π. In modern terms, Archimedes has discovered that $3.1408 < \pi < 3.1429$, so he has fixed the first two decimal places of π.

This defines **calculation 3, Archimedes bounds π**. Archimedes has essentially introduced what we now call the physical constant π (although he did not explicitly say that or use the symbol π), and he has shown how to find bounds on its value. The idea of finding upper and lower bounds for a quantity when an exact calculation is not available is an important one in mathematics and science. Archimedes's method of calculation is also highly innovative. All in all, this is a remarkable result, clearly worthy of inclusion in any list of important calculations.

Archimedes completes the story in "Measurement of the Circle" with

Proposition 2. The area of the circle is to the square on its diameter as 11 to 14.[4]

This result ($A = {}^{11}\!/_{14}\,d^2$) follows simply by using the 22/7 upper bound in equation (2.4). Thus Archimedes has shown how to relate a circle's circumference C and its area A to its diameter d.

2.2.3 Some Computational Details

Archimedes made his calculation by comparing the circumference of a circle with the perimeters of polygons fitting inside and outside the circle as in the examples shown in figure 2.2. For the four-sided polygon (a square) the reader may easily derive the bounds $2\sqrt{2} < \pi < 4$ or $2.83 < \pi < 4$. For the six-sided polygon, or hexagon, the bounds are $3 < \pi < 2\sqrt{3}$ or $3 < \pi < 3.3642$. These are not very useful results, and it took the genius of Archimedes to see a clever way to go beyond them.

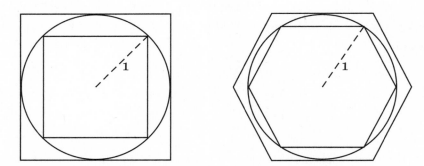

Figure 2.2. Approximating a circle using polygons with four sides (a square) and six sides (a hexagon). A circle with radius 1 has a circumference equal to 2π.
Figure created by Annabelle Boag.

Archimedes began with hexagons, which give the quite-poor bounds mentioned above. Archimedes realized that he needed to use polygons with many more sides to generate accurate results. His idea was to go in steps in which the number of sides would be doubled each time. He increased the number of sides from 6 to 12, then to 24, to 48, and finally to 96, at which point he was satisfied with the results, which are given in equations (2.5) and (2.6).

What makes Archimedes's calculation so brilliant is that he showed how to use the answer for a polygon with n sides to generate the answer for a polygon with $2n$ sides; it was not necessary to go right back to the beginning each time. The angles in a polygon with $2n$ sides are half those in a polygon with n sides and this fact, together with Euclid's:

> Book VI, Proposition 3. If an angle of a triangle is bisected and the straight line cutting the angle cut the base also, the segments of the base will have the same ratio as the remaining sides of the triangle,[5]

allowed Archimedes to find the length of a side of a $2n$-sided polygon simply from the length of the n-sided one. Along the way he had to calculate square roots, and the manipulations used are equally impressive. (The reader wishing to follow the complete details should see the original working in the books by Heath and Dijksterhuis or the more modern version as set out by Stein.)

Recall that Archimedes started with a hexagon whose side length involves $\sqrt{3}$. To proceed beyond that he needed a measure of that square root and he came to

$$\frac{265}{153} < \sqrt{3} < \frac{1351}{780}.$$

These are extremely accurate bounds on $\sqrt{3}$, and their origin is discussed in the references just given.

The innovative conceptual design and the method of working produced by Archimedes surely tell us that his calculation of circle properties is truly one of the great calculations.

Another of Archimedes's innovative calculations will be referred to in chapter 3.

2.2.4 Archimedes Has Begun the Race

With his calculation, Archimedes started what we can view as a race to determine the ratio of a circle's circumference to its diameter—a number which today we call π, following the suggestion of William Jones in 1706. The following highly selective list gives an idea of the progress over the centuries. (The results have been converted to decimal form for comparison with $\pi = 3.14159265358 \ldots$)

Date	Author	Approximation for π
150	Ptolemy	377/120 or 3.14167
263	Liu Hui	3.14159
480	Tsu C'ung Chi	3.1415926
800	Al-Khowarizimi	3.1416
1593	Viete	to 9 decimal places
1596	Van Ceulen	to 20 decimal places
1615	Van Ceulen	to 35 decimal places
1700	Machin	to 100 decimal places
1844	Strassnitzky & Dase	to 200 decimal places
1949	Smith & Wrench	to 1,120 decimal places
1973	Guilloud & Bouyer	to 1,001,000 decimal places
1989	Kanada & Tamura	to 1,071,741,799 decimal places

(You should consult the bible for this subject—the source book edited by Berggren, Borwein, and Borwein—for details of these and many other calculations, and in fact, for anything you might ever want to know about π!)

The above (highly selective) list tells you that the search for π was an international affair. Archimedes's polygon method was carried to extremes by people like Van Ceulen whose result was reputedly carved onto his tombstone. Calculations by Machin and others represent the start of a new era in which mathematical analysis and calculus were used to present new expressions for calculating π, which has led to a whole research field providing super-convergent series. I just cannot leave out two early beautiful results (even though they are hopeless from a computational point of view):

$$\text{John Wallis in 1655: } \pi = 2 \times \frac{2 \times 2 \times 4 \times 4 \times 6 \times 6 \times 8 \times 8 \ldots}{3 \times 3 \times 5 \times 5 \times 7 \times 7 \times 9 \times 9 \ldots}$$

$$\text{James Gregory in 1670: } \pi = 4\left(1 - \frac{1}{3} + \frac{1}{5} - \frac{1}{7} + \frac{1}{9} - \frac{1}{11} + \ldots\right).$$

You will notice that early results gave π as a fraction like 22/7. This raises the question of whether there is actually an exact expression for π of the form n/m where n and m are both integers; that is, whether π is a rational number. A negative answer was proven by Lambert in 1768, and in 1882 Lindeman proved that π is not just irrational, it is also transcendental. That also solved an old problem, taking us back to Euclid and his method of dealing with areas in terms of equivalent figures. Using the algebraic theory related to geometric constructions, Lindeman's result tells us that Euclid could never give a ruler-and-compass construction of a square with the same area as a given circle. Thus the old squaring-of-the-circle problem was finally put to rest (although cranks and eccentrics have never given up the quest).

Look at almost any area of science, and you will find π in the equations and formulas. Few constants could be more important or better known, and Archimedes told us how to evaluate it. In so doing he introduced the idea of bounding a quantity and demonstrated the power of iterative methods, two innovative and outstanding contributions to mathematics.

2.3 A DIFFICULT DECISION

I suspect that many readers will be asking: Is that it? Nothing more from the ancient world? I confess that it has been difficult to stop at those two calculations, but for me they are supreme examples of what early mathematicians were achieving. I feel particularly bad about leaving out material from the ancient Chinese and Indian mathematicians, but I did set a difficult limit of fifty calculations for the whole book. (I suggest that readers interested in this period consult the books listed in the bibliography; I particularly recommend the book by Joseph, who has done much to restore balance in the history of mathematics.)

Chapter 3

STEPS INTO MODERN MATHEMATICS

*in which we look at four steps on the way to
our modern mathematical world.*

Mathematics from the ancient world was preserved by the Arabs, and after hundreds of years, it became available to mathematicians in what we now call "the West." The following four examples show how that process began a progression of thinking and ideas that finally led to the mathematics we know today.

3.1 CALCULATION 4: FIBONACCI SHOWS US HOW TO DO IT

The calculations in this book are presented using our familiar representation of numbers, which uses the symbols 0, 1, 2, 3, 4, 5, 6, 7, 8, 9, and arithmetic to base 10. Thus 463 means 4 hundreds, 6 tens, and 3 units. Early mathematicians used other approaches, and in section 2.1 we saw the Mesopotamians using base 60. Archimedes used the very clumsy Greek representation of numbers involving a system of letters to indicate numbers (see Dijksterhuis, chapter 3, section 0.6). Equally awkward to use were the Roman numerals, which today we use in a decorative way as when we write MMXIV for the year 2014. Our modern number system has its origins in India and, like much other ancient knowledge, it was recorded and nurtured in the Arab world. After a long period of comparative inaction, the growth of learning and science in most of Europe only began again when that knowledge was translated from Arabic into Latin, ready for western scholars to make use of it. For arithmetic and the art of calculation, we can assign two vitally important dates: 1202 and 1228.

The Prologue of the great book *Liber Abbaci* opens with:

> Here begins the Book of Calculation
> Composed by Leonardo Pisano, Family Bonaci
> In the Year 1202.

It is this book, *Liber Abbaci, The Book of Calculation*, that changed the way people did their mathematical work. The second edition, published in 1228, is still in use today.

3.1.1 The Author

The exact details of Leonardo's birth are not known, but as his name indicates, he grew up in Pisa. His father was involved in commerce, and naturally, Leonardo was schooled in basic mathematics and its uses. However, the event that ultimately changed the course of history was his father's posting to Bugia (in modern Algeria) as a diplomatic representative of Pisa and contact for trade with Africa. The young Leonardo went to live with his father in Bugia and also traveled around the region. So it was that he came to learn about and appreciate the accounting and mathematical methods used by the Arabs and trading nations around the Mediterranean and far beyond. Obviously Leonardo was a talented young man because he absorbed all he could and became the foremost mathematician of his time. The publication of his *Liber Abbaci* marks a turning point in the spread of mathematics and its applications.

Today Leonardo Pisano is better known as Fibonacci, a nickname introduced in 1838 by the historian Guillaume Libri. (I will use Fibonacci here, as most people are more familiar with that name. His story is given in the recent biography by Keith Devlin and in the older, little book by Joseph Gies and Frances Gies.)

3.1.2 The *Liber Abbaci*

We are lucky to have the modern translation of *Liber Abbaci* by Laurence Sigler. Three things become apparent when looking at the book. First, it is an enormous work, about 600 pages in Sigler's translation. This is not just a simple textbook or mathematical record; it is a compendium of mathematical results. Second, the title translates as "Book of Calculation," and the arithmetic is much as we know it today; this is not a book about the abacus or other calculating devices. Third, this is a book for use by teachers and merchants wishing to learn the new approach to numbers. *Liber Abbaci* contains an enormous number of examples, both serious and entertaining.

Liber Abbaci comprises fifteen chapters, with the first seven devoted to the basics of numbers and arithmetic. Five chapters present applications that are of importance for merchants, accountants, and so on. The final three chapters deal with mathematical techniques, with the usual array of innovative examples.

Chapter 1 opens with the key to the whole book and the progress in mathematics that it produced:

> The nine Indian figures are
>
> $$9\ 8\ 7\ 6\ 5\ 4\ 3\ 2\ 1.$$
>
> With these nine figures and the sign 0 which the Arabs call zephyr, any number whatsoever is written, as is demonstrated below. A number is a sum of units, or a collection of units, and through the addition of them the numbers increase by steps without end.

He goes on to explain about tens, hundreds, and so on. There are examples of how those awkward Roman numerals could be neatly handled:

> write 2023 for MMXXIII and write 4301 for MMMMCCCI.

The chapter concludes with simple addition and multiplication tables, just as we learn them in primary school today. Tables for manipulating fractions are given in later chapters.

Chapters 2–7 develop arithmetic using those "Indian figures." For example, chapter 2 teaches us how to multiply, and like any good teacher, Fibonacci builds up a series of ever more complicated examples such as:

$$7 \times 308 = 2156, \quad 70 \times 81 = 5670, \quad 123 \times 456 = 56088,$$
$$\text{and finally } 12345678 \times 87654321 = 1082152022374638.$$

A measure of Fibonacci's thoroughness is gained by noting that he suggests checking answers using the old technique called "casting out nines" and (of course) detailed examples are given.

The next four chapters show how to use the new arithmetic in the banking, trading, and the business world:

8. On Finding the Value of Merchandise by the Principal Method
9. On the Barter of Merchandise and Similar Things
10. On Companies and their Members
11. On the Alloying of Monies

Next is the large (187 pages in Sigler's translation) and untitled chapter 12. Fibonacci illustrates many mathematical procedures (including the sum of an arithmetic progression and the sum of a series of squares) using a large collection of what today we might call recreational problems (see below, sections 3.1.3 and 3.1.4).

Finally there are three chapters on more advanced mathematical ideas. Fibonacci shows how to find square and cube roots using approximation techniques, including the idea of building up a series of ever more accurate answers (as we saw Archimedes doing for bounds on π). There is a great deal of algebra here (and in earlier chapters) but the modern reader may struggle to appreciate that as it is given in the rhetorical form, where words but no symbols are used, which was prevalent before Viete introduced our modern symbolic form in 1591.

In summary, in *Liber Abbaci*, Fibonacci introduced part of the world to arithmetic easily performed using Hindu numerals and set out a breath-taking array of examples of all kinds.

3.1.3 A Recreational Mathematics Example

"Purse problems" have a long history, and Fibonacci included his own versions. Here is one from chapter 12:

> Two men who had denari found a purse with denari in it; thus found, the first man said to the second, if I take these denari of the purse then with the denari I have I shall have three times as many as you have. Alternately, the other man responded, and if I shall have the denari of the purse with my denari, then I shall have four times as many as you have. It is sought how many denari each has, and how many denari they found in the purse.

Today we solve this problem using a little linear algebra. Let the two men have x_1 and x_2 denari, and the purse contain p denari. Then the problem tells us that

$$p + x_1 = 3x_2 \quad \text{and} \quad p + x_2 = 4x_1,$$

which we easily solve to get $x_1 = \left(\frac{4}{11}\right)p$ and $x_2 = \left(\frac{5}{11}\right)p$. In the second of his approaches to the problem, Fibonacci does some similar algebra except that for him, in the rhetorical form, an unknown is "the thing" (whereas we would use a symbol like x). We notice that there are three unknowns (p, x_1, and x_2), and the two conditions in the problem only allow us to find two unknowns, which are therefore given in terms of the third. The next step is to notice (as Fibonacci effectively does) that we are only interested in whole numbers, so we will say there are 11 denari in the purse ($p = 11$), and then the men have 4 and 5 denari. (Of course, we could use any multiple of 11 as p.) Fibonacci is cleverly using the fact that his problem requires a solution in whole numbers not fractions.

3.1.4 But What about the Rabbits?

Of all the 600 pages in *Liber Abbaci*, the content of one particular page in the chapter on recreational mathematics has become extremely famous:

How Many Pairs of Rabbits are Created by One Pair in One Year?

A certain man had one pair of rabbits together in a certain enclosed place, and one wishes to know how many are created from the pair in one year when it is the nature of them in a single month to bear another pair, and in the second month those born to bear also.

Fibonacci shows how to proceed month by month and create what we now call the Fibonacci numbers 1, 2, 3, 5, 8, 13, 21, 34, 55, 89, 144, 233, and 377. Thus there will be 377 pairs of rabbits in twelve months. (I return to the modeling of population growth in chapter 12.) Fibonacci is demonstrating the explosive nature of the growth of rabbit populations. Obviously the early settlers in Australia had not read *Liber Abbaci* because in 1859 they released twenty-four rabbits into the wild, and the resulting plague continues to cause trouble for farmers to this very day!

The Fibonacci numbers are now known to occur in a great range of biological problems, and through one simple example, he created a whole research field. Pure mathematicians have discovered many interesting properties of the whole sequence of Fibonacci numbers.

3.1.5 Evaluation

Fibonacci wrote other books, including *Liber Quadratorum* in which he demonstrated his talents as a mathematician by presenting a series of increasingly involved propositions about the squares of numbers (now a venerable topic in number theory). However, in *Liber Abbaci*, Fibonacci is gathering together a magnificent array of mathematical techniques and examples, which he collected from a variety of sources and which he supplements with his own ideas and methods. It is a very approachable, comprehensive treatise, and it changed the way people did mathematics for the next three hundred years. For this reason, I record **calculation 4, Fibonacci's presentation in *Liber Abbaci***. A fitting summary is given by Keith Devlin in his biography of Fibonacci:

The greatness of Liber Abbaci is due to its quality, its comprehensive nature, and its timeliness. It was good, it provided merchants, bankers, business people, and scholars with everything they need to know about the new arithmetic methods, and it was the first to do so. Though there were to be many smaller, derivative texts that would explain practical arithmetic, it would be almost three hundred years before a book of comparable depth and comprehensiveness would be written.[1]

3.2 CALCULATION 5 TO MAKE LIFE EASIER

Which would you rather do: multiply 512 by 128, or add 9 and 7? No doubt you settled for the addition. If the numbers were larger, it would make your choice even more emphatic—or would it? It may be that you respond by saying the question seems a little silly and you do not much care; after all it is only the choice of which button to press on a calculator. However, in the days before calculating machines were readily available, addition was comparatively simple, but a complicated multiplication would be an agonizing and time consuming task. So it was in 1614 that John Napier came to write:

> Seeing as there is nothing (right well beloved students in the Mathematics) that is so troublesome to Mathematical practice, nor that doth more molest and hinder Calculators, than the Multiplications, Divisions, square and cubical Extractions of great numbers, which besides the tedious expense of time, are for the most part subject to many slippery errors. I began therefore to consider in my mind, by what certain and ready Art I might remove those hinderances.[2]

The stage was set for the invention of an invaluable new mathematical technique.

3.2.1 The Basic Idea

Take a look at table 3.1.

n	2^n	n	2^n
0	1	9	512
1	2	10	1,024
2	4	11	2,048
3	8	12	4,096
4	16	13	8,192
5	32	14	16,384
6	64	15	32,768
7	128	16	65,536
8	256	17	131,072

Table 3.1. Numbers n in an arithmetic progression
and a corresponding geometric progression.

The columns giving the n values form an arithmetic progression while the corresponding 2^n columns form a geometric progression (just as we saw in section 1.2). In the first case, additions are involved as we move down the columns, whereas it is multiplications in the second case. I also remind you about (or introduce you to) the exponent laws:

$$2^n \times 2^m = 2^{(n+m)}, \qquad 2^n \div 2^m = 2^{(n-m)}, \qquad 2^{2m} = 2^m \times 2^m \quad \text{so} \quad \sqrt{2^{2m}} = 2^m .$$

From the table we can see that 512×128 is the same as $2^9 \times 2^7$ and the exponent law tells us that is $2^{9+7} = 2^{16}$. Consulting the table gives 2^{16} as 65,536 so we conclude that $512 \times 128 = 65,536$. Thus we have seen that writing numbers as 2 to some power allows us to use addition of exponents to carry out tedious multiplications. You can show how divisions are carried out by finding that $512/128$ is $2^{9-7} = 2^2 = 4$. Because 65,636 is 2^{16} its square root must be $2^8 = 256$.

While we have seen a wonderfully simple way to carry out complicated calculations just using additions or subtractions, the table of numbers on which we can use those ideas is extremely limited. However, for any given number y it is possible to find an exponent x so that

$y = 2^x$ and x is called the logarithm of y, written as $\log_2(y) = x$.

Naturally, x is no longer a simple integer for most cases. Log is short for logarithm and the subscript 2 on the log symbol is to remind us that we are using 2 as the base for our calculations. In fact we can use any number as the base (obviously requiring new tables like 3.1), and we shall see that base 10 has become a standard. Table 3.1 is just a table of logarithms with base 2; in the heading we could replace 2^n with y and n by $\log_2(y)$. For example, for 256 the table gives 8 as $\log_2(256)$.

The exponent laws can now be written as laws for manipulating logarithms:

$$\log(y_1 \times y_2) = \log(y_1) + \log(y_2)$$

$$\log(y_1 \div y_2) = \log(y_1) - \log(y_2)$$

$$\log(\sqrt{y}) = \left(\tfrac{1}{2}\right)\log(y).$$

Using table 3.1 we find

$\log(512 \times 128) = \log(512) + \log(128) = 9 + 7 = 16 = \log(65{,}536)$

and so we deduce that $512 \times 128 = 65{,}536$.

3.2.2 John Napier

The basic ideas about logarithms and the recognition of their value were first explained by John Napier (1550–1617). He was the eighth Laird of Merchiston and after 1608, lived in Merchiston Castle near Edinburgh. He was a very active and inventive man whose writings covered many subjects including agriculture, engineering, and military matters. He also fathered twelve children. (See the book by Calinger for further details.)

Napier was a staunch Scottish Presbyterian and enemy of Catholics, and his best-selling work, *A Plaine Discovery of the Whole Revelation of Saint John: Set Down in Two Treatises*, ran through twenty-one editions in English and had many translations.

John Napier seems to be an unlikely mathematical innovator, but the quote given in the first paragraph of section 3.2 shows that he well understood the perils of calculation; his great discovery was a method for enormously reducing those perils. Napier introduced logarithms into mathematics, but not in the form given above. Napier hit upon the essential point—the use of the correspondence between the terms in an arithmetic series and those in a geometric series—but he dealt with it in terms of the motion of points along two lines, in one case with uniform speed (so distance traveled forms an arithmetic progression) and in the other with variable speed (to give effects like a geometric progression). (See the book by Goldstine and the articles by Bruce, Carslaw, and Jagger for details.) Using clever mathematical methods and a great deal of time (which only someone like a nobleman might have), Napier produced tables of logarithms, and his acclaimed *Mirifici Logarithmorum Canonis Descriptio* was published in 1614. According to historian Ronald Calinger "his [Napier's] logarithms were the cardinal computational accomplishment of the Reformation."[3]

3.2.3 Henry Briggs

Naturally, many people became interested in logarithms, but the major figure in their development was Henry Briggs (1561–1631). He was the first professor of geometry in the new Gresham College in London and later held the same position at Merton College in Oxford. Briggs recognized the magnitude of Napier's achievements, and, in 1615, he made the arduous journey to Edinburgh to consult Napier. Briggs stayed for a month, and the two men came up with the ideas that led to logarithms as we know them today.

One major change saw the adoption of base ten for future logarithms. (The construction method used by Napier results in a complicated structure as explained in the references cited above.) Briggs used a number of ingenious mathematical methods to calculate logarithms—see Goldstine and Bruce's aptly named paper "The Agony and the Ecstasy—The Development of Logarithms by Henry Briggs." In his 1624 *Arithmetica Logarithmetica*, Briggs gave the base ten logarithms for all the integers from 1 to 20,000 and from 90,000 to 100,000 with instructions on how to find the other obviously missing cases. He gave the results to fourteen-figure accuracy! The second edition, published in 1828 with the help of the Dutch mathematician Adriaan Vlacq, covered all integers up to 100,000 but reduced to ten-figure accuracy. Briggs also published extensive tables of sines and tangents of angles and their logarithms in his *Arithmetica Logarithmic*, sometimes working to fifteen-figure accuracy. These were tables needed in many applications, and it was centuries before there was any improvement over them.

These tables revolutionized the business of calculation and were soon seized upon by scientists and engineers, particularly those needing to manipulate expressions involving trigonometry.

3.2.4 Impact

The invention of logarithms and tables of them were a boon for people such as astronomers (like Kepler) and surveyors. (The cited books by Calinger and Jagger tell the story of the widespread use and importance of logarithms.) The situation was famously summed up by the most eminent of all French scientists, Pierre-Simon Laplace, when he said that logarithms, "by shortening the labors, doubled the life of the astronomers."[4] Napier himself seems to have been in no doubt about the value of his invention because his *Mirifici Logarithmorum Canonis Descriptio* opens with a poem which, translated from the original Latin, reads:

> This book is small if you consider just the words,
> but if you consider its use, Dear Reader, this book is huge.
> Study it and you will learn that you owe as much to this little book
> as to a thousand large volumes.[5]

As a person who was at school and college in the pre-electronic-calculator era, I know just what Napier meant! I have no hesitation in adding to the list **calculation 5, production of tables of logarithms**.

Napier invented a wooden device ("Napier's bones") for mechanically using logarithms. William Oughtred and others made advances in that line, and eventually it led to the "modern" slide-rule, which was an essential part of the toolkit of engineers until the era of electronic calculators. (For more information, see the book by M. R. Williams, and it is also worth seeking out the old *Handbook of the Napier Tercentenary Exhibition* edited by E. M. Horsburgh.)

Finally, logarithms give a fine example of the way mathematicians can sometimes transform the nature of a problem using a mathematical technique; in this case, multiplications turned into additions by moving into the world of logarithms. Similar advances were made later by introducing Fourier transforms and Laplace transforms.

3.3 THE MASTER GIVES US CALCULATION 6

In mathematics, we often create a sequence of numbers or answers. We have seen Malthus calculating values for a population and its resources as they develop over a set of time intervals, Archimedes finding a sequence of ever-better approximations for π, and Fibonacci showing us how to find the number of pairs of rabbits as their population explodes. The next step in some cases is to sum the sequence of terms, and we call that a mathematical series. The sum is denoted by s with a subscript counting how many terms a_i are to be included:

$$s_4 = a_1 + a_2 + a_3 + a_4 \text{ and generally } s_n = a_1 + a_2 + a_3 + \ldots + a_n.$$

For example, using terms as in Malthus's arithmetic progression we find

$$1 + 2 + 3 + 4 + 5 = 15.$$

In his *Liber Abbaci*, Fibonacci gave the general formula for summing an arithmetic progression which starts with the number a and increases by d each term:

$$s_n = a + (a+d) + (a+2d) + ... + (a+[n-1]d) = \left(\tfrac{1}{2}\right)n(2a + [n-1]d), \quad (3.1)$$

and Fibonacci wrote $s_n = \left(\tfrac{1}{2}\right) \times$ the number of terms $\times$ the sum of the first and last terms.

Fibonacci also gave the sum of squares:

$$1^2 + 2^2 + 3^2 + 4^2 + ... + n^2 = \left(\tfrac{1}{6}\right)n(n+1)(2n+1).$$

Malthus used a geometric progression, and the sum for a general problem of this type (starting with a and multiplying by r to get each new term) is also well known:

$$s_n = a + ar + ar^2 + ar^3 + ... + ar^{n-1} = \frac{a(1-r^n)}{(1-r)} \quad \text{or} \quad \frac{a(r^n - 1)}{(r-1)}. \quad (3.2)$$

A problem on the ancient Egyptian Rhind papyrus (see section 2.1.4) called for this sum with $a = 1$ and $r = 7$ taken to 5 terms and got the answer 2,801. The problem is often repeated in various similar forms; for us it is the old puzzle "as I was going to St. Ives . . .," while for Fibonacci in the *Liber Abbaci*, it was "seven old men go to Rome . . ."

But now, a word of caution: the above results are special, and it is not always easy to find a formula for the sum in a series. For example, finding s_n for the so called "harmonic series,"

$$S_n = \tfrac{1}{1} + \tfrac{1}{2} + \tfrac{1}{3} + \tfrac{1}{4} + \tfrac{1}{5} + ... + \tfrac{1}{n} \quad (3.3)$$

is notoriously difficult.

3.3.1 The Next Level

Mathematics often involves the infinite, and it is natural to ask what happens when we let the series continue on forever. We can ask for the value of s_n as n becomes as large as we like ("in the limit as n approaches infinity"). Clearly, for an arithmetic progression such as in equation (3.1), the answer becomes infinitely large and we say the series diverges. For a geometric progression as in equation (3.2), the answer depends on the multiplier r; when $r > 1$, the series also diverges, but for $r < 1$, it has the finite limit $a/(1-r)$ (since in that case r^n approaches zero for large n.) A series with a finite answer is said to converge. An attractive old example is this geometric series (with $r = \frac{1}{2}$):

$$1 + \tfrac{1}{2} + \tfrac{1}{4} + \tfrac{1}{8} + \tfrac{1}{16} + \ldots = 2.$$

In general, it is not easy to discover whether an infinite series diverges or converges, and finding the actual sum in the convergent case is usually even harder. But before going further, perhaps I should respond to readers who are asking why we would bother with such questions at all. The answer (apart from natural curiosity and accepting a challenge), is that infinite series are extremely important in the development of mathematics and its applications. An ancient example appears in Archimedes's wonderful calculation of the area enclosed by a parabola and a line cutting it (see the books by Dijksterhuis and Stein in the chapter 2 bibliography). To get his result, Archimedes had to prove that the infinite geometric series with $a = 1$ and $r = \frac{1}{4}$ has a sum of 4/3, that is

$$1 + \tfrac{1}{4} + \tfrac{1}{16} + \tfrac{1}{64} + \tfrac{1}{256} + \ldots = \tfrac{4}{3}.$$

The development of calculus lead to the use of infinite series (especially by people like Isaac Newton) for representing various functions. For example:

$$\sin(x) \;=\; x - \frac{x^3}{3!} + \frac{x^5}{5!} - \frac{x^7}{7!} + \ldots$$

$$\log_e(1+x) \;=\; x - \frac{x^2}{2} + \frac{x^3}{3} - \frac{x^4}{4} + \frac{x^5}{5} - \ldots \tag{3.4}$$

The second expression was very useful for calculating logarithms.

3.3.2 The Problem and the Master

In all of this, one problem became particularly famous and one man gave its solution—along with a great many other results.

The harmonic series in equation (3.3) when extended to an infinite number of terms was proved to diverge by Oresme as long ago as 1350. But what if the terms reduced in size even faster as in

$$s_\infty \;=\; 1 + \frac{1}{2^2} + \frac{1}{3^2} + \frac{1}{4^2} + \frac{1}{5^2} + \ldots \tag{3.4}$$

This series does converge; even though there are an infinite number of terms, they reduce in size so quickly that the total sum is finite. But finding the value of the sum caused great difficulties. Because this problem was publicized by Jacob Bernoulli, who lived in the Swiss city of Basel, it became known as the famous Basel problem.

The man to tackle such problems was Leonard Euler (1707–1783). He was actually born in Basel, but in later life he served in academies in Russia and Germany. Euler was surely the most prolific mathematician of all time, and few branches of pure mathematics do not show signs of his pioneering inputs. A poll of readers of the journal *Mathematical Intelligencer* produced a ranking of the most beautiful results in mathematics;[6] the top two and the fifth ranked are results from Euler. He was also an applied mathematician and scientist of great repute with interests spanning many areas. It was Euler who took Newton's mechanics and turned them into the form we use today. He fathered thirteen children and still found time to write textbooks in calculus and algebra. Euler became perhaps the first great popularizer of science. His *Lettres á une Princesse d'Allemagne* (*Letters to a German Princess*) became what today we would call a publishing sensation, going through many editions and being translated into German, English, Dutch, Swedish, Italian, Danish, and Spanish. They remain an interesting and informative introduction to science even today. (The book by Dunham gives a good introduction to Euler and his work, and the paper by Kline deals specifically with infinite series.)

When asked about how to tackle mathematical problems, Laplace (the "French Newton") responded: "Read Euler; He is our master in everything."[7] So it is with series; Euler was truly the master of the topic and his *Introduction to Analysis of the Infinite* is simply breathtaking in its scope and inventiveness.

Euler solved the Basel problem and gave the answer

$$1 + \frac{1}{2^2} + \frac{1}{3^2} + \frac{1}{4^2} + \frac{1}{5^2} + \ldots = \frac{\pi^2}{6}.$$

For me this is an amazing, neat, and beautiful result (and mathematicians tend to agree as it was fifth in the poll mentioned earlier). Who would have expected that ubiquitous π to turn up here! (Actually, as part of his proof, Euler used expressions like those in equation (3.4) and any link to sine will suggest that π might be involved—see Dunham or Euler's own book for details.)

Euler went further, and he could give results for other even powers such as

$$1 + \frac{1}{2^4} + \frac{1}{3^4} + \frac{1}{4^4} + \frac{1}{5^4} + \ldots = \frac{\pi^4}{90},$$

$$1 + \frac{1}{2^6} + \frac{1}{3^6} + \frac{1}{4^6} + \frac{1}{5^6} + \ldots = \frac{\pi^6}{945}.$$

These results by the master are my **calculation 6, Euler solves the Basel problem**.

It should not be thought that these results are mere curiosities (otherwise they would not be in my list of great calculations). Euler developed new approaches to series and the mathematics of functions expressed as power series, which have affected the whole progress of mathematics. One or two examples are given in the next section for those wishing to see some details.

3.3.3 A Technical Comment

Euler showed how to generalize the Basel problem and then how to link it to the theory of prime numbers (more of which in the next section). Euler considered what we now call the zeta function (since it is denoted using the Greek letter ζ):

$$\zeta(m) \;=\; 1 + \frac{1}{2^m} + \frac{1}{3^m} + \frac{1}{4^m} + \frac{1}{5^m} + \;\ldots. \qquad (3.5)$$

He was able to study this series as a function of m; for example, he proved that it converges for all $m > 1$. In one of those steps which only the master might make, Euler linked the zeta function to the prime numbers:

$$\zeta(m) \;=\; \left(\frac{1}{1-1/2^m}\right) \times \left(\frac{1}{1-1/3^m}\right) \times \left(\frac{1}{1-1/5^m}\right) \times \left(\frac{1}{1-1/7^m}\right) \times \ldots \quad (3.6)$$

and so the infinite series in equation (3.5) is shown to be equivalent to an infinite product involving just the prime numbers. Euler showed how problems about numbers could be tackled using the methods of analysis. I will return to this in the next section.

3.4 GAUSS, HEROIC CALCULATIONS, AND THE PRIME NUMBER THEOREM

Johann Carl Friedrich Gauss (1777–1855) was six years old when Euler died, and he was already recognized as a child prodigy in mathematics and calculations. Gauss went on to contribute broadly in mathematics and physics and is usually put on the same level as giants like Archimedes and Newton. Gauss gave us the famous hierarchy: "Mathematics is the Queen of the Sciences and Number Theory is the Queen of Mathematics."[8] So it is fitting that Gauss should provide the next great calculation in the field of number theory.

The most special and important of all numbers are the primes (those numbers not divisible by any other number). Already the ancient Greeks knew that any number (and here we are talking about the positive integers) can be written as a unique product of prime numbers. For example, 126 = 2 × 3 × 3 × 7. This result is called the fundamental theorem of arithmetic. Thus the primes are the building blocks for all numbers and a key to understanding arithmetic and number theory results. An obvious question is: How many prime numbers are there? The answer, as Euclid showed in his *Elements*, is that the list of primes goes on forever; there is no largest prime, or as some people like to say, there are an infinite number of primes. (This is a little disappointing when we compare the physical case of atoms, which are all built up using just three particles: the electron, the proton, and the neutron.)

The next questions come when we look at the primes in more detail. Here are all the primes less than 150:

2	3	5	7	11	13	17	19		
23	29	31	37	41	43	47	53	59	
61	67	71	73	79	83	89	97		
101	103	107	109	113	127	131	137	139	149

Looking at the first primes suggests that they come in pairs: 3 and 5, 5 and 7, 11 and 13, 17 and 19. Such pairs of prime numbers that differ by 2 are known as twin primes. However, the next twins do not appear until 41 and 43, then 71 and 73. There are more twins, like 101 and 103, but no obvious pattern emerges. There is probably an infinite number of twin primes—nobody knows for sure. That example is typical; it is hard to see any pattern in the prime numbers.

Another approach is to seek formulas or equations that produce sets of primes. For example, Euler discovered that the polynomial $x^2 - x + 41$ gives prime numbers when x is set equal to 1, 2, 3, . . . 40. (See the books by Wells, and Conway and Guy for more on such things.) Lots of results have been found, but no single overall method for generating the primes is known.

If we step back further, we can ask the seemingly simple question: How many primes are there less than or equal to a given number x? The result is called $\pi(x)$. (It seems unfortunate that the Greek letter π is used here given that it is so widely used in terms of circle properties. The notation $\pi(x)$ was introduced by Edmund Landau in a 1909 book about prime numbers.) Counting up the primes listed in the above table gives that $\pi(20) = 8$, $\pi(71) = 20$ and $\pi(150) = 35$. It is not at all obvious how $\pi(x)$ depends on x. We need to see how $\pi(x)$ changes as x is increased to larger and larger values. We need to do some calculations!

It is here that we see a perfect counterexample to T. H. Huxley's famous statement:

> Mathematics is that study which knows nothing of observation, nothing of experiment, nothing of induction, nothing of causation.[9]

Such an idea was strongly refuted at the 1869 meeting of the British Association. If they needed a dramatic example they could have used the extensive calculations made to discover the factorization of a range of numbers. When no factors are found, then of course the number is a prime and can be added to the list. The story of these calculations is most remarkable. Here is the 1980 summary by the eminent mathematical historian Howard Eves:

> Extensive factor tables are invaluable for research on prime numbers. Such a table for all numbers up to 24,000 was published by J. H. Rahn in 1659 . . . In 1668 John Pell of England extended this table up to 100,000. As a result of appeals by the German mathematician J. H. Lambert, an extensive and ill-fated factor table was computed by a Viennese schoolmaster named Antonio Felkel. The first volume of Felkel's computations,

giving factors of numbers up to 408,000, was published in 1776 at the expense of the Austrian imperial treasury. But, as there were very few subscribers to the volume, the treasury recalled almost the entire edition and converted the paper into cartridges to be used in a war for killing Turks. In the nineteenth century, the combined efforts of Chernac, Burck-hardt, Crelle, Glaisher, and the lightning mental calculator Dase, led to a factor table covering numbers up to 10,000,000 and published in ten volumes. The greatest achievement of this sort, however, is the table cal-culated by J. P. Kulik (1773–1863), of the University of Prague. His as yet unpublished manuscript is the result of a twenty-year hobby, and covers all numbers up to 100,000,000. The best available factor table is that of the American mathematician D. N. Lehmer (1867–1938); it is a clev-erly assembled one-volume table covering numbers up to 10,000,000. Lehmer has pointed out that Kulik's table contains errors.[10]

Lehmer himself wrote a history of these calculations (see the bibliography).

3.4.1 Using the Data

Gauss was involved with mathematical tables throughout his life. At four-teen he was given mathematics books, including tables of logarithms, by the Duke of Brunswick who recognized his talents and supported him during the early part of his life. Around the age of sixteen, Gauss wrote down his conjecture for the behavior of $\pi(x)$ after studying tables of primes. Other people, like Legendre, also came close to giving the full story.

Based on the evidence, Gauss concluded that the nature of $\pi(x)$ is sum-marized in the prime number theorem:

as x becomes large, $\pi(x)$ asymptotically approaches $x/\log(x)$,

thus $\dfrac{\pi(x)}{x/\log(x)}$ approaches 1 as x tends to infinity.

Two technical points: the log here is the natural log with base e; Gauss later gave a more accurate result using the logarithmic integral $Li(x)$:

as x becomes large, $\pi\big(x\big)$ asymptotically approaches $Li(x) = \int_2^x 1/\log(x)\ dx$.

Since $Li(x)$ is asymptotically equal to $x/\log(x)$ we can use either formula; it is just that for smaller values of x, the integral is more accurate. For example, if $x = 1{,}000{,}000{,}000$, the number of primes $\pi(x)$ is 50,847,478; the prime number theorem gives 48,254,942, and Gauss's $Li(x)$ gives 50,849,235. If x is taken ever larger, the values of $\pi(x)$, $x/\log(x)$, and $Li(x)$ become closer and closer. (Details of Gauss's personal calculations and samples of the tables he produced are given in the papers by Goldstein and Tschinkel, who also reproduce a letter from Gauss to his student Encke containing some interesting historical points.)

The prime number theorem at last reveals some sort of order in the set of primes. It is a beautiful example of the way calculations provide insight and inspiration, and for that reason it is my important **calculation 7**.

3.4.2 Beyond the Calculations

The link between analysis and number theory pioneered by Euler eventually led to independent analytical proofs of the prime number theorem in 1896 by J. Hadamard and C. J. de la Vallée-Poussin.

Research into the primes has produced an enormous number of fascinating results. (See the book by Wells for a simple introduction.) For example, Euler showed that if we keep only the terms involving primes in the famous harmonic series, equation (3.3), the series still diverges. However, if we use only the twin primes, Viggio Bruns showed in 1919 that

$$\tfrac{1}{3} + \tfrac{1}{5} + \tfrac{1}{5} + \tfrac{1}{7} + \tfrac{1}{11} + \tfrac{1}{13} + \tfrac{1}{17} + \tfrac{1}{19} + \tfrac{1}{41} + \tfrac{1}{43} + \ldots = 1.9021605\ldots$$

Thus this infinite series does converge, and by 2002 the sum was known to even greater accuracy: 1.902160583104. Such are the charms and delightful results of prime number theory.

It is impossible not to mention the most famous unsolved problem involving the primes. In 1742, Christian Goldbach wrote to Euler asking about numbers and the sum of primes. The result is the famous Goldbach Conjecture: every even number, except 2, can be written as the sum of two primes. For example, $16 = 5 + 11$ and $160 = 59 + 101$. The conjecture has been verified by calculations with numbers up to the staggeringly large 4×10^{14}, but to this time, no analytical proof has been found to back up the calculations (and if you find one, there is a \$1 million prize you can collect).

Prime numbers offer challenges and lots of fun for professional and amateur mathematicians alike. The factorization problem has also become important in a practical way. The difficulty of factoring a very large number into a product of primes is the basis of the RSA algorithm for public key encryption of data for transmission in military and business contexts. (Wells give more details and references.)

3.5 WHAT IS LEFT OUT

The choice of calculations for this chapter is almost limitless, and I am sure every reader will have some favorite work that I have excluded in my desire to limit the size of this book. Some of you may wonder how I could ignore Ramanujan, for instance. Here are just five topics that ended up in my extensive near-misses category.

1. Pascal's triangle contains the numbers used in binomial theorem expansions and is also used in combinatorial problems and in the theory of probability. Like the primes, it leads to many fascinating mathematical properties and uses (see the book by Edwards). This is also an example of a topic that was explored by early Chinese mathematicians (see the books by Martzloff and by Li Yan and Du Shiran in the chapter 2 bibliography).

2. Rafael Bombelli (about 1526–1573) solved a cubic equation using a standard procedure, but as part of his working, he bravely manip-

ulated expressions involving the square root of minus one. This led people to believe such things could enter mathematics, and so began the use of complex numbers. It is almost impossible to imagine modern pure or applied mathematics without the use of complex numbers.

3. Linear equations and sets of such equations are ubiquitous in mathematics, and few techniques in mathematics could be more famous than Gaussian elimination as a solution method for them. Gauss made some very important calculations using this method. This is another example where early Chinese mathematicians also showed how to systematically tackle relevant problems (see references to Martzloff, and Li Yan and Du Shiran, as mentioned above).

4. The classification of groups resulted in some mammoth calculations carried out by many mathematicians. The book by Marcus du Sautoy tells the wonderful story, including how group classification involves integers eighteen digits long.

5. As a quirky example, I offer Benford's law, also known as the first digit phenomenon, which states: in many collections of numbers, including tables of various kinds, the first digit is most commonly 1; if not 1, then 2; and so on in a logarithmic distribution. There is something quite captivating about this and it even has applications. To go further, try the paper by Hill and Berger.

Chapter 4

OUR WORLD

*in which we see how mathematics helped to answer
questions about our world and sometimes stirred up major
controversies in so doing; learn more about how calculations
are linked to experiments and the difficulties involved;
and see some of the different approaches used
when making calculations.*

As science gradually developed man began to ask questions about
the world in which he lived, something that continues today. I
have chosen four examples for discussion in this chapter, and a
fifth appears in the next chapter. As always, there are many more that I
might have included, and at the end I will discuss some of them and make
a few general comments about calculations in this area. The four examples will illustrate the theory-experiment link and show different facets of
physical calculation work.

4.1 HOW BIG IS THE EARTH?

We naturally appreciate the size of things around us and start to measure
them in terms of convenient units—millimeters for small things, meters
for larger everyday objects, and kilometers for longer journeys. Interestingly, the very first letter Leonard Euler wrote for his wonderful *Letters to
a German Princess* is titled *On Magnitude*. Measuring the size of things is
one of the most basic tasks of science. Once it was accepted that the earth
can be taken as a sphere, it was natural to ask how big it is. Sailors knew
about the curved horizon and vanishing landmarks as they sailed out to

sea, and various schemes were used to estimate the size of planet Earth. Aristotle (of course) gives a value, although not a very accurate one.

Eratosthenes devised a different approach: he showed how data from an experiment could be used in a simple calculation to relate the circumference of Earth to a given known distance in Egypt. Eratosthenes was born around 285 BCE in Cyrene, a Greek city in the part of North Africa we know as Libya, and lived to be about eighty. He was educated in Athens, and in 245 BCE, he moved to Egypt to become head librarian at the illustrious Musaeum in Alexandria. He was what today we would call a polymath and made contributions across many fields of science, mathematics, and the arts. It is a tragedy that none of his original work survives today, but there are good descriptions of his life and achievements written by his contemporaries and other ancient scholars. The book by Nicastro gives a readable introduction to the man and his work.

4.1.1 Eratosthenes's Calculation

Eratosthenes knew that at the summer solstice, the sun at its zenith in Syênê was directly overhead as indicated (reputedly) by the fact that it shone down to the bottom of wells. If the angle $\theta°$ to the vertical made by the sun at the same time at Alexandria was measured, the geometry shown in figure 4.1 could be used.

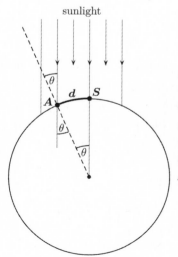

Figure 4.1. Geometry for the sun's rays striking the earth at Alexandria (A) and Syênê (S), which are a distance d apart. *Figure created by Annabelle Boag.*

We assume Alexandria and Syênê are a distance d apart on the same north-south meridian, and then, by a simple proportion argument, we get

$$\frac{\text{distance between cities, } d}{\text{measured angle, } \theta\,^\circ} = \frac{\text{circumference of earth, } C}{360^\circ},$$

$$C = \frac{360}{\theta}\, d. \tag{4.1}$$

The measurement technique for obtaining θ is shown in figure 4.2. Measuring the shadow cast by a vertical rod gives the required angle through the fact that the shadow length is proportional to $tan(\theta)$. If the rod is at the center of a spherical dish then the length of the shadow in the dish is directly proportional to the angle θ.

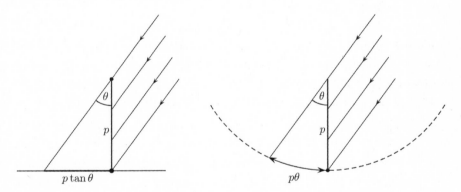

Figure 4.2. Measurement of angle θ using the shadow cast by a vertical rod of length p on a plane or a spherical surface. *Figure created by Annabelle Boag.*

Eratosthenes found that θ was one fiftieth of a whole circle (or 7.2°), and the distance d between the cities was 5,040 stades. Equation 4.1 gives the earth's circumference C as 50 × 5,040 or 252,000 stades. This may be translated as 39,690 kilometers, which is remarkably close to the modern figure for the polar circumference: 40,009 kilometers. The mathematics may be trivially simple, but this calculation at last gave a believable estimate for the size of the earth. It is my **calculation 8, Eratosthenes measures the earth.**

4.1.2 Analysis

Because this is such a simple and transparent calculation, it provides a good example for learning about various factors that should be considered when a calculation is used to describe a physical situation. Perhaps surprisingly, there are many issues involved.

Any physical theory will rest on certain **basic assumptions**. In this case, Eratosthenes assumes a spherical Earth (or at least a great circle for the circumference under consideration). Newton later showed that the earth is not a sphere, and that is confirmed by accurate surveying. However, little error is introduced here. Eratosthenes must also assume that the rays of the sun are parallel as they strike the earth, which is not too unreasonable. Finally he must assume that Alexandria and Syênê do actually lie north-south on a great circle passing through the poles. In fact, their longitudes differ by about 3°, so a small error is made.

Any calculation of a physical quantity will need **input data**, which may be a mixture of physical constants and data from measurements. In this case, values for θ and d are required. We cannot be certain exactly how θ was measured, and the fact that the sun is of finite size means there would be some fuzziness in the shadow being observed. According to Dutka (see bibliography) θ should be 1/50.6, not 1/50, times the value 360 for the full circle. We do not know the precision with which Eratosthenes lined up the rod so that it was vertical and pointing to the center of the earth as assumed in figure 4.2. An error is thus introduced, but we should note that the magnitude that Eratosthenes calculates for C is still substantially correct. The other piece of information required is the value of d. There is much speculation about how d might have been measured—see the references to Nicastro and Dutka. Again some error was obviously introduced, but it is not sufficient to invalidate Eratosthenes's conclusion.

For any calculation, we need to consider the **interpretation of the result**. In this case it is clear that a meaningful estimate of the earth's circumference has been obtained. The real problem here is to be sure what the ancient unit of a stade actually referred to. This is a complicated matter, and I refer you to Nicastro and Dutka for a very detailed analysis of the possibilities.

In a modern calculation, there is often an **assessment of likely error** given alongside the final result. In this way, the various uncertainties in the method and the input data can be taken into account to give error bounds for the answer.

Despite all of the above points, I believe that Eratosthenes came up with a very good value for the size of the earth, and his work is worthy of inclusion on any possible list of great calculations.

4.2 WHAT IS THE MASS THE EARTH?

If we know the size of the earth and make some assessment of its density, we may find an approximate value for the mass of the earth. This approach is linked to knowledge of the constitution of the earth which will be considered in section 4.4. I will return to the mass of the earth in the next chapter where a quite different approach will be discussed.

4.3 HOW OLD IS THE EARTH?

All cultures have their myths and creation stories, some of which suggest how long ago the earth was created. For some people, including Aristotle, no age was involved; the world was eternal. In the nineteenth century, particularly in Britain, a conflict arose as a number of groups of people were interested in estimating the age of the earth, and diverse figures were being suggested. At that time, the question was still quite open: Was the earth a few thousand years old? Hundreds of thousands of years old? A few million years old? Or even many millions of years old? This was a time of great progress and upheaval in the sciences, and knowing the age of the earth was of major importance. (The article by Badash gives a good brief history of the subject.)

For Christians, the Bible may be taken as the book giving all necessary knowledge; the Bible may be assumed to give the literal truth, a belief still held by many people even today. The creation of the earth is described in

Genesis, and the chronology of later events and generations may be used to work back to a creation date. Several people worked on the details (see chapter 2 in the book by Weintraub), but the most famous is Bishop James Ussher (1581–1656). He worked with great precision to come to the view that

> In the beginning God created Heaven and Earth, Genesis. 1, v. 1, which beginning of time, according to our chronologie, fell upon the entrance of the night preceding the twenty third day of October in the year of the Julian Calendar, 710 [equivalent to 4004 BCE].[1]

Thus the fundamental religious view suggests an age of thousands of years for the earth. Although many people would dismiss Ussher's estimate as ludicrous, we should note that it actually involved a very long and detailed analysis of the events described in the Bible, and an intricate calculation is required to trace back through the generations to the first persons placed by God upon the earth. A Christian fundamentalist might even suggest this to be an innovative approach worthy of consideration as a great calculation. It should also be noted that figures like Kepler and Newton also considered a few thousands of years to be a valid estimate for the age of the earth.

The nineteenth century saw tremendous progress in geology and biology, the two merging in the study of fossils found in different strata of rocks. Sir Charles Lyell's great *Principles of Geology, An Attempt to Explain the Former Changes of the Earth's Surface by References to Causes Now in Operation* was published in three volumes in 1830–1833. Charles Darwin took the first volume of Lyell's treatise on his round-the-world voyage on the Beagle, and it played a part in his work leading to his *Origin of Species* in 1859. This was a period of theorizing on a grand scale. Some geologists were convinced that the earth was many hundreds of millions of years old. Darwin himself, in his *Origin*, came up with a figure of about 300 million years for the time taken for water to have cut out the valley of the Weald near his home. (This calculation was not endorsed by many people, and Darwin removed it from the second edition of his book.) He also required a very great age for his theory of evolution based on natural selection to be valid. The scene was set for a major conflict with one of the great physicists of the period.

4.3.1 Enter Lord Kelvin

William Thomson (1824–1907) was, together with James Clerk Maxwell, the leading figure in nineteenth-century British physics. Thomson was made Lord Kelvin in 1892 and is now universally known just as Kelvin. His contributions to science and life in Britain were enormous, and it is no surprise to find Smith and Wise needing over 800 pages for their biography of him (see bibliography).

Kelvin was one of the founders of thermodynamics and explained its influence on the classical theory of heat and mechanical processes. In particular, he understood the conservation of energy and the manner in which all physical actions involve friction and processes by which available energy gradually runs down. Thus it was natural that he turned his attention to our source of energy for life, the sun. He tried to understand how the sun produced energy and calculate a possible lifetime for it, suggesting in 1862 that the sun could not have illuminated the earth for more than 100,000,000 years, or 500,000,000 at the outside. (I will return to this topic in chapter 11.) Clearly this had implications for the age of the earth.

In 1862, Kelvin published his famous paper *On the Secular Cooling of the Earth*. (This paper is freely available today and is a joy to read.) He begins his paper with this challenge to geologists and their trust in "uniformitarianism":

> For eighteen years it has pressed on my mind, that essential principles of Thermo-dynamics have been overlooked by those geologists who uncompromisingly oppose all paroxysmal hypotheses, and maintain not only that we have examples now before us, on the earth, of all the different actions by which its crust has been modified in geological history, but that these actions have never, or have not on the whole, been more violent in past time than they are at present.[2]

Kelvin suggests that the earth has been cooling from some initial state, and the theory of heat as developed by Fourier may be used to calculate how long the cooling has been taking place to give the present conditions. Others before Kelvin had wondered about the cooling of the earth (Newton mentions a cooling ball of iron in his *Principia*), but Kelvin had the theoretical tools to produce meaningful time estimates.

4.3.2 Kelvin's Calculation

Kelvin assumes that "the earth is a warm chemically inert body cooling" from molten rock assumed "to be at 7000° Fahr." To carry out the calculations, he uses Fourier's equation for the temperature T at time t and position x:

$$\frac{\partial T}{\partial t} = \kappa \frac{\partial^2 T}{\partial x^2}. \tag{4.2}$$

Fourier's theory tell us that if we are given the temperature at points on an extended body at time $t = 0$, we may solve equation (4.2) to find the temperature $T(x, t)$ at any future (or past) time as heat conduction takes place.

Kelvin studied cooling in a planar geometry (arguing later that it may be used to give suitable estimates for the spherical case), so that only the one spatial coordinate x is used. The parameter κ denotes the conductivity of the cooling material. Kelvin uses a solution of equation (4.2) which gives the state of a cooling system as a function of time by taking a boundary condition making the surface held at a constant temperature. He gives an age for the earth of 98,000,000 years but states that "the consolidation cannot have taken place less than 20,000,000 years ago, or we should have more underground heat than we actually have, nor more than 400,000,000 years ago."

Thus Kelvin satisfies no one: the biblical estimates are made to look ridiculous, and the times required by the theories in geology and biology appear to be far too large. Kelvin himself was a practicing Christian. Ivan Ruddock describes how, in his vote of thanks at a somewhat anti-Darwin lecture, and in subsequent correspondence in the *Times* (London), Kelvin revealed his belief that "science positively affirmed Creative Power."[3] Perhaps the fact that his age for the earth did not cover the period required by evolution was not such a worry for Kelvin.

Kelvin's calculation was an application of physical principles on a grand scale, and it stirred up an extensive debate among the various types of scientists of his time. There is no doubt that Kelvin's calculation led to an examination of the underlying assumptions and methods in many areas of science, and thereby it changed the direction of science in a major way. For that reason I choose it as **calculation 9, Kelvin and the age of the earth**.

4.3.3 Analyzing Kelvin's Calculation

Kelvin's work on the cooling of the earth is what today we would call a model calculation: the details of a physical situation are simplified or approximated in such a way that a manageable mathematical theory may be applied. Kelvin, in his 1862 paper, makes it quite clear that "the solution thus expressed and illustrated applies, for a certain time, without sensible error, to the case of a solid sphere, primitively heated to a uniform temperature, and suddenly exposed to any superficial action, which for ever after keeps the surface at some constant temperature." He is also clear that he needs a value for the conductivity parameter κ, and he discusses how he chooses possibilities from certain experimental evidence. He does also discuss some limitations of his model (more of which in a moment).

These circumstances were appreciated by those engaged in the long debate about physics, geology, evolution, and the state of the earth. A wonderful example is provided by T. H. Huxley ("Darwin's bulldog") when he said:

> Mathematics may be compared to a mill of exquisite workmanship, which grinds you stuff of any degree of fineness; but, nevertheless, what you get out depends upon what you put in; and as the grandest mill in the world will not extract wheat-flour from peascod, so pages of formulae will not get a definite result out of loose data.[4]

Huxley is setting out general ideas which should be kept in mind when considering any of the calculations described in this book.

Kelvin's model simplifications were analyzed by John Perry, who was at one time Kelvin's assistant before going on to professorships in Tokyo and London. Perry published several papers in *Nature* (see bibliography) commenting on Kelvin's assumptions and what they implied for his calculations. In his January and February 1895 papers, he pointed out that κ would vary in the extreme conditions inside the earth, and he offered some sample calculations showing that the ages given by Kelvin could be increased by a factor of more than a hundred. Kelvin responded in an 1895 *Nature* paper defending an age of 24 million years. Perry reviewed that again in his April paper where he also discussed the fact that Kelvin's model ignores other processes like convection in the interior of the earth. Perry concluded that Kelvin's ages were much too short. We can only conclude that Kelvin made far too sweeping simplifications and it is not easy to model the extremely complex body that is our earth, something we return to below in section 4.4.5. (For a full discussion of debates around Kelvin's work see the paper by England, Molnar, and Richter and the book by Smith and Wise.)

Kelvin may not have got to the 4.6 billion years accepted today as the age of the earth, but through his calculation, he did produce a revolution in thinking about science and the ways it is applied on those large scales becoming so prominent in nineteenth-century science.

4.3.4 Rutherford's Wonderful Story

It is impossible to leave this topic without touching on one of science's great stories. Kelvin assumed that there was no source of heat within the earth (recall that it was "a warm chemically inert body cooling") and so only a calculation of cooling from some initial state was required. However, it was in Kelvin's lifetime that radioactivity was discovered, and thus a source of heat within the earth was identified. In 1904, Ernest Rutherford gave an address at the Royal Institution, and there in the audience was the eighty-year-old Lord Kelvin! Here is Rutherford's famous description of the event:

> I came into the room, which was half dark, and presently spotted Lord Kelvin in the audience and realized that I was in for trouble in the last part of the speech dealing with the age of the Earth, where my views conflicted with his. To my relief he fell fast asleep but as I came to the

important point, I saw the old bird sit up, open an eye and cock a baleful glance at me! Then sudden inspiration came, and I said Lord Kelvin had limited the age of the Earth, *provided no new source of heat was discovered.* That prophetic utterance refers to what we are now considering tonight, radium! Behold! The old boy beamed at me![5]

Heat generated by radioactivity inside the earth represents another flaw in Kelvin's calculations. However, he did not believe radioactivity would greatly affect his calculations (see Smith and Wise, chapter 17), and indeed it does not appear that such a source of heat is too important (see the England, Molnar, and Richter paper). Ironically, it is methods based on the physics of radioactivity that are used to give good estimates for the age of Earth.

4.4 WHAT IS INSIDE THE EARTH?

We may readily explore the surface of the earth, climb up mountains, and go down into valleys, gorges, and mines. That does not take us very far inside the earth and the deepest hole drilled so far is around 12 km deep compared with the earth's radius of 6,360 km. Naturally people have wondered what is inside the earth, and it has been a subject for myths and science fiction. It might seem to be an impossible quest, but in fact we do know a great deal about the inside of our earth. In the broadest terms, there is a crust (which we can explore to a certain extent) on top of the mantle, which extends down to 2,890 km; there is an outer core in the 2,890–5,150 km region; and finally there is a largely iron inner core.

In the nineteenth century, scientists could use mathematics to investigate the connection between the properties of the interior of the earth and its shape, rotation, orbit, and tidal variations. These were and remain very indirect methods needing many factors and physical phenomena to be taken into account. This was the realm of the great classical physicists like Lord Kelvin. But as the twentieth century dawned, a new and potentially more accurate method was coming onto the scene.

4.4.1 Earthquakes and Their Detection

An earthquake sends out disturbances in the earth that may be detected all over the globe. In the nineteenth century, many instruments (seismographs) were developed for recording the effects of an earthquake at sites remote from its center (see Oldroyd, chapter 10 for details). As more and more observations were made, it was inferred that the disturbances traveled as waves, and since they moved through the earth, perhaps those observed waves held the clue to the constitution of the interior.

Two types of waves propagate through a solid body like the earth. P-waves involve compressions and dilations; they are pressure waves (like sound waves), and they describe traveling longitudinal displacements in the material through which they are propagating. S-waves involve displacements perpendicular to the direction of motion (like the waves on an elastic string); they are shear waves. The speeds of these waves depend on the elastic properties of the transmission medium, and a comprehensive mathematical theory has been built up over the last two centuries (see Bullen and Bolt for examples).

P-waves travel faster than S-waves. We should also note the simple, but vitally important fact that shear waves like S-waves cannot propagate in liquids. There are also waves that run along the surface of solids. They too may involve longitudinal or transverse effects, and they are named after their investigators, Rayleigh and Love. (See Bullen and Bolt. The book by Bolt gives a good, simple introduction to seismic waves.)

The path of wave fronts may be traced out using rays. It is the spread in the times taken as waves follow particular ray paths from the earthquake that is the key information provided by the seismic-wave detectors. Because P-waves travel fastest, they arrive first, hence the notation primary (P) and secondary (S) waves.

Like other waves, the P- and S-waves reflect at boundaries between different media (like light at a mirror or sound waves forming an echo); refract or bend as they move from one region into a different one (like light rays at the air-water interface); and follow curved paths if the elastic properties of the medium in which they are propagating vary continuously (like light rays as they give mirages in the variable refractive index lower atmosphere).

There is one additional important property of seismic waves as they meet a surface: a mixture of waves is generated. Thus if a P-wave meets a surface there will be transmitted and reflected P-waves, but also newly generated transmitted and reflected S-waves. Because P- and S-waves have different speeds, the generated S-waves do not make the same angle with the surface as the P-waves do. This is a complication that turns out to be a key to interpreting certain seismograph results.

One final point about P- and S-waves: as they reflect and refract, they will take different paths from the source to the observation point. There will be regions on the earth where one or another type of wave will not be observed because of their propagation characteristics. Those "shadow regions" are very important pieces of information for the theorist to explain.

Figure 4.3 shows seismic rays within the earth. At this stage, you may prefer just to glance at it to get an idea of the complications involved as rays reflect, refract, and convert to a different form. Although many people wrestled with the problem of seismic waves and what they tell us, I have chosen three people to illustrate the gradual progress to the picture we have today. (The paper by Brush gives a more complete story.)

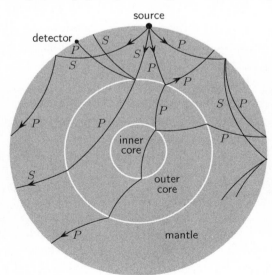

Figure 4.3. Seismic rays within the earth. P- and S-waves are represented, and the various changes at interfaces can be observed. *Figure created by Annabelle Boag. Drawn using data from B. A. Bolt's 2004 book (see bibliography), which should be consulted for full details.*

4.4.2 Richard Oldham and the Use of Seismic Waves

Richard Oldham (1858–1936) left England in 1879 to join the Geological Survey of India, working there until ill health forced his return to England in 1903. He became familiar with the rapidly accumulating seismic data, and in 1900, published "On the Propagation of Earthquake Motion to Great Distances." Here, Oldham was able to gather together data from many events and put it into a graphical form that clearly revealed groupings and trends in transmission times. His graph shows data forming three distinct curves. He was able to identify P- and S-waves, give information about their speeds, and also identify a third phase wave. Then he made the prophetic statement:

> If the curves drawn on the diagram represent the true time curves, it should be possible to deduce from them the relation between the variation of velocity of transmission and depth below the surface.[6]

After much work, Oldham published his famous 1906 paper "The Constitution of the Interior of the Earth, as Revealed by Earthquakes." He opens by summarizing the limited knowledge of the earth then available and notes that "the central substance of the earth has been supposed to be fiery, fluid, solid, gaseous in turn, till geologists have turned in despair from the subject." He goes on to say that the days of speculation may be over as seismograph data becomes available and interpreted. Oldham then boldly makes his celebrated statement:

> Just as the spectroscope opened up a new astronomy by enabling the astronomer to determine some of the constituents of which distant stars are composed, so the seismograph, recording the unfelt motion of distant earthquakes, enables us to see into the earth and determine its nature with as great a certainty, up to a certain point, as if we could drive a tunnel through it and take samples of the matter passed through.[7]

To "see into the earth" in this way requires large, detailed calculations of ray paths and disturbance arrival times for different models of the earth. Eventually Oldham became confident enough in his calculations to make statements about P- and S-wave speeds in the earth and to come to his sur-prising conclusion:

> From the considerations detailed in the foregoing pages, I conclude that the interior of the earth, after the outermost crust of heterogeneous rock is passed, consists in a uniform material, capable of transmitting wave-motion of two different types at different rates of propagation: that this material undergoes no material changes of physical character to a depth of about six-tenths of the radius, each change as takes place being gradual and probably accounted for sufficiently by the increase in pres-sure; and that the central four-tenths of the radius are occupied by matter possessing radically-different physical properties, inasmuch as the rate of propagation of the first phase is but slightly reduced, while the second-phase waves are either not transmitted at all, or, more probably, trans-mitted at about half the rate which prevails in the outer shell.

A diagram showing how rays occur in a model Earth with a core was given by Oldham and is shown in figure 4.4. Richard Oldham calculated the wave speeds necessary to fit the seismograph observations and made one of science's great discoveries: the earth has a distinct inner core.

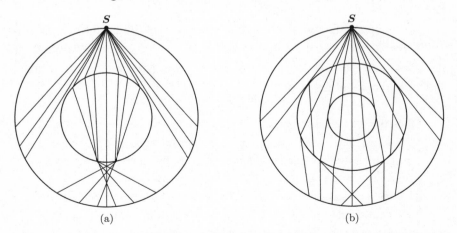

Figure 4.4. Seismic-ray paths in an earth with a core according to (a) Oldham and (b) Lehmann. *S* is the source or origin of the rays. Uniform media are assumed so that the rays follow straight lines. *Figure created by Annabelle Boag. Redrawn from the papers of Oldham and Lehmann (see bibliography).*

4.4.3 Inge Lehmann Refines the Core Model

Following Oldham's 1906 paper, there was investigation and speculation by many scientists about the nature of the earth's core. Oldham himself, in a 1913 *Nature* paper, suggested that fluids or gases could be involved. Sir Harold Jeffreys, one of the greats of seismology, concluded in 1926 that the core is truly fluid. Remembering that S-waves cannot propagate in fluids, this explains the comments by Oldham on their absence in his 1906 paper.

The next major step came in the work of Inge Lehmann (1888–1993). Lehmann was Danish, studied mathematics at the Universities of Copenhagen and Cambridge, and then had a career in seismology. Her personal and scientific story may be read in her 1987 reminiscences "Seismology in the Days of Old" and in the review by Kölbl-Ebert. It is unfortunate that this exceptional representative of women in science is so little known. Two quotes reported by Kölbl-Ebert sum up Lehmann's (and sadly many other women's) experiences. Of her schooling, Lehmann said that "no difference between the intellect of boys and girls was recognized, a fact that brought some disappointment later in life when I had to recognize that this was not the general attitude," and "you should know how many incompetent men I had to compete with—in vain."[8]

Lehmann studied the seismograph records from the 1928 Mexican and 1929 New Zealand earthquakes. In particular she looked at the strength of the waves and their shadow zones over the Earth. After many calculations, she took a momentous step:

> I then placed a smaller core inside the first core and let the velocity in it be larger so that a reflection would occur when the rays through the larger core met it. After a choice of the velocities in the inner core was made, a time curve was obtained, part of which appeared in the interval where there had not been any rays before. The existence of a small solid core in the innermost part of the earth was seen to result in waves emerging at distances where it had not been possible to predict their presence.[9]

Sample rays as given by Lehmann are shown in figure 4.4. (The reader wishing to get a better understanding of this should see the very clear and detailed figure 2 in the Kölbl-Ebert paper. The paper by Rousseau is a detailed but relatively simple introduction to Lehmann's discovery of the inner core.)

So it was that Oldham's prediction became fact, and, by analyzing the seismic records, the crust/mantle/fluid-outer-core/solid-inner-core model was discovered Although much refined and improved, it remains as today's accepted picture of the earth.

4.4.4 Mohorovičić and His Tantalizing Discontinuity

The Croatian Andrija Mohorovičić (1857–1936) was a pioneering seismologist who is best known for his 1910 discovery about the boundary between the earth's crust and mantle. In October 1909 there was an earthquake in Croatia with an epicenter about 40 km south-east of Zagreb. Mohorovičić had access to local seismographic data, which he plotted out and concluded:

> [it] cannot be expressed by only one curve, there are two curves: one beginning in the epicenter reaching distances up to 700 km., certainly not beyond 800 km. Second, lower curve, begins certainly at 400km, but it is possible that it has already started at 300 km.[10]

Mohorovičić was convinced that the two curves related to the same type of wave. (See the papers by Herak, and by Jarchow and Thompson for the curves and further details.) Mohorovičić interpreted the data using the model shown in figure 4.5. After completing his calculations he concluded that at a depth of 54 km there was a distinct boundary between the crust and the mantle, with wave speeds changing from 5.68 km/sec to 7.75 km/sec as the boundary is crossed. Thus there is a discontinuity, and seismic rays are reflected by it to give ray paths as shown in figure 4.5. This explained the two curves and the shadow zone that Mohorovičić observed in the 1909 earthquake seismographic data.

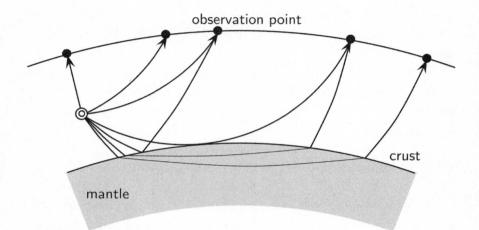

Figure 4.5. Mohorovičić's diagram showing seismic rays reflected at the crust-mantle discontinuity. *Figure created by Annabelle Boag. Drawn from information in the paper by Herak (see bibliography).*

The Mohorovičić discontinuity, now known as the Moho, is defined by Jarchow and Thompson as "that level in the Earth where the compressional wave velocity increases rapidly or discontinuously to a value between 7.6 and 8.6 km/sec."[11] The depth of the Moho varies from 5 km to 8 km for crust below deep ocean basins and from 20 km to 70 km for continental crust. It is the comparatively small depth below the ocean that makes Mohorovičić's such a tantalizing result; can we drill down to the Moho and examine the physical changes in detail? (For a recent report on progress I refer you to "Drilling to Earth's Mantle" by Umino, Nealson, and Wood.)

4.4.5 Discussion

This topic is a beautiful example of the way calculation may be used to turn data into information about the physical world. In this case, the method is essential since we are exploring the details of regions of Earth that are not directly accessible. As a good candidate for the label "great," I list **calculation 10, seismic rays reveal the earth's interior**.

From a technical point of view, while the calculations of ray paths and times may be relatively simple given the properties of the material through which the waves propagate, reversing the problem to one of finding the material properties leading to a given set of ray paths can be particularly troublesome. The first problem (find the rays given the material) is called the direct problem; the second problem (find the material properties) is called the inverse problem. Inverse problems are notoriously difficult; they can be unstable, depend very sensitively on input data, and raise questions about the uniqueness of the solution.

The work discussed here has been extended enormously and is still providing forefront research problems (see the recent reviews by Buffett and Olson). The properties of the core depend on the behavior of matter at extreme pressures and temperatures, and these are now active areas in solid-state physics and chemistry. (For a short discussion and an example about the structure of iron under extreme conditions, see the 2010 Physics Update on "Iron's Structure at Earth's Core.") On the darker side, the use of seismic waves for detecting nuclear explosions has been a valuable tool for monitoring nuclear test ban treaties (the modern network of 337 recording facilities is described by Auer and Prior).

Finally, we can look back to Lord Kelvin's calculation on the cooling of the earth and appreciate the complexities of the system he was attempting to model. Perry was correct when he began to identify all sorts of possibilities that invalidate Kelvin's simple model calculation.

4.5 MOTION ON A SPINNING GLOBE

The earth is a large sphere rotating about a north-south axis to give us our twenty-four-hour day with alternating daylight and nighttime. That rotation combined with our knowledge of the circumference of the earth leads us to conclude that we are moving at over 1,500 kilometers per hour. That enormous speed was a stumbling block for the early acceptance of the rotating Earth concept. Why are the effects of motion at such a great speed not apparent? Why aren't birds and clouds left behind? Why don't our hats fly off as they might do on even a galloping horse? These are profound questions, and they require an equally profound advance in science to answer them. If we have to pick one great early pioneer in this work, it clearly must be Galileo, and it is he who is responsible for the next calculation in my list.

As discussed in the first chapter, mathematics plays an essential part in science. Perhaps the most famous—and certainly one of the most beautifully expressed—statements of this guiding idea came from Galileo:

> Philosophy is written in that great book which ever lies before our eyes—I mean the Universe—but we cannot understand it if we do not first learn the language and grasp the symbols in which it is written. This book is written in the mathematical language, and the symbols are triangles, circles and other geometrical figures, without whose help it is humanly impossible to comprehend a single word of it, and without which one wanders in vain through a dark labyrinth.[12]

The work discussed below is taken from Galileo's *Two New Sciences*, which is readily available today. It is easy enough to read, although it does become rather tedious as it is written in terms of a long, and at times rambling, debate among the three protagonists spread over four days and an "added day." Also, the mathematics used by Galileo is couched in a geometric form quite unfamiliar to present-day readers. Before coming to the major calculation, we need to see how Galileo introduces the types of motion involved.

4.5.1 Basic Concepts

The Third Day of Galileo's debate is called *On Local Motion*, and Galileo tells us he is introducing "a brand new science concerning a very old subject."[13] He begins by defining constant speed or uniform motion:

> Equal or uniform motion I understand to be that of which the parts run through in any equal times whatever are equal to one another.

He can now make the important point that "motion in the horizontal plane is equable, as there is no cause of acceleration or retardation."

That leads him to state what we might call the idea of inertia and thus be reminded of Newton's first law:

> It may also be noted that whatever degree of speed is found in the moveable, this by its nature indelibly impressed on it when external causes of acceleration or retardation are removed, which occurs only on the horizontal plane; . . . From this it likewise follows that motion on the horizontal is also eternal, since if it is indeed equable it is not weakened or remitted, much less removed.[14]

The old Aristotelian idea of a continuing cause for motion is removed; once in motion in the horizontal plane, an object continues that way. Thus a hat leaving your head retains its motion and is not left behind at a rate of over a thousand km/hr. Objects on the surface of the Earth all move in the same "equable" manner.

Galileo also needs to describe motion not with constant speed, but with constant acceleration:

> I say that motion is equably or uniformly accelerated which, abandoning rest, adds on to itself equal momenta of swiftness in equal times.[15]

The speed increases by the same amount in any equal time intervals.

Galileo deduces properties of uniformly accelerated motion by using a geometric representation of the quantities involved. (This idea goes back to Nicole Oresme (1325–1382), amongst others, but Galileo was not one for giving credit to his predecessors.) In figure 4.6 (a), the line CD represents the distance traveled; in the accompanying diagram, the time taken is measured along the line AB, and the speed at any given time is measured by the line perpendicular to AB, thus forming the line AE. The final speed is given by BE. The distance traveled is the area between the lines AB and AE (what today we refer to as the area under the speed-versus-time curve), so the total distance traveled is given by the area of the triangle AEB. Simple geometry tells us that triangle AEB and rectangle $AGFB$ have the same area. Thus by comparing areas in his diagram, Galileo can now give his first result:

The time in which a certain space is traversed by a moveable in uniformly accelerated movement from rest is equal to the time in which the same space would be traversed by the same moveable carried in uniform motion whose degree of speed is one-half the maximum and final degree of speed [*EB*] of the previous, uniformly accelerated motion.

This is sometimes called the average speed rule, or the Merton rule, and was known well before Galileo's time.

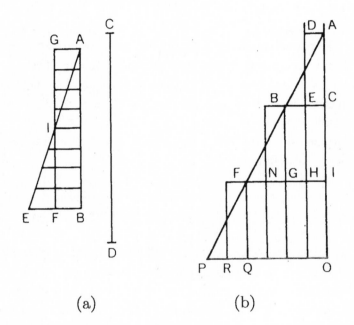

(a)   (b)

Figure 4.6. Galileo's diagrams for describing uniformly accelerated motion. (a) shows what happens in the time interval *AB*. (b) shows how equal time intervals *AC*, *CI*, and *IO* give increasingly different distances traversed. *From Galileo Galilei,* Discourses and Mathematical Demonstrations Relating to Two New Sciences *(*Discorsi e dimostrazioni matematiche, intorno à due scienze*) (1638).*

Similar geometric reasoning concerning the areas in the diagram in figure 4.6 (b) corresponding to the equal time intervals *AC*, *CI*, and *IO* leads Galileo to his famous result that in a series of equal time intervals, the distance traveled "will be to one another as are the odd numbers from unity, that is, as 1, 3, 5, 7, . . ." The total distances traveled are of course like 1, 1+3 = 4, 1+3+5 = 9, 1+3+5 +7 = 16, . . ., "that is as the squares of the times."

These examples illustrate Galileo's geometric way of working; the geometric figure with its line lengths and areas replaces the algebra in a modern calculation.

4.5.2 Projectile Motion

Day Four of the dialogues is called *On Projectile Motion*, and here Galileo settles an old question: how to determine the path of a projectile and discover its form. Galileo has talked about equable motion in a horizontal plane, and now he must describe motion in a vertical plane. In his book he explains that

> a heavy body has from nature an intrinsic principle of moving toward the center of heavy objects (that is, of our terrestrial globe) with a continually accelerated movement, and always equally accelerated.[16]

Hence Galileo can use his results for uniformly accelerated motion for movement toward the center of the earth. He can now deal with a projectile by considering together, but independently, its horizontal and vertical motions. He has one of the debaters clearly make this point:

> Assuming that the transverse motion is always kept equable, and that the natural downward motion likewise maintains its tenor of always accelerating according to the squared ratio of times; and also that such motions, or their speeds, in mixing together, do not alter, disturb, or impede one another.[17]

Because it involves equable or uniform motion, a horizontal distance traveled is a measure of the time taken. Galileo thus traces out a motion with a vertical distance traveled in a given time proportional to the square of the horizontal distance traveled in that same time. Comparing this with the geometry of a parabola, he reaches his monumental conclusion:

When a projectile is carried in motion compounded from equable horizontal and from naturally accelerated downward motions, it describes a
semiparabolic line in its movement.[18]

The great question is answered: using his geometric method of calculation Galileo has shown that projectiles trace out parabolic paths.

4.5.3 The Maximum Range Property

Using his result, Galileo can now settle another question: At what angle
should a projectile be launched to give maximum range? Galileo uses his
geometric method to show that

the maximum projection, or amplitude of semiparabola (or whole
parabola) will be that corresponding to the elevation of half a right
angle.[19]

Thus the launch angle should be 45° to get the maximum range. Galileo
gives a table of numerical results giving the range for varying launch
angles at one degree intervals, and he limits the extent of the table by using
another of his results: the range reduces by the same amount if the launch
angle is increased or decreased by a certain number of degrees around 45°.
Thus the range is the same for launch angles 48° and 42°, or for angles 35°
and 55°, or for angles 30° and 60°.

Galileo was clearly proud of his achievement, and he has one debater
saying the demonstration "is full of marvel and delight." He has also demonstrated how "the knowledge of one single effect acquired through its
causes opens the mind to the understanding and certainty of other effects
without need of recourse to experiments." Undoubtedly a worthy addition
to my list is **calculation 11, Galileo describes projectile motion**.

4.5.4 Afterward

Galileo recognizes that his model for projectile motion does not include air
resistance. A theory including air resistance had to wait for Newton and the

idea of forces in dynamics. Except for a very special case, the force due to air resistance couples the horizontal and vertical motions and destroys the parabolic nature of the trajectory. Eventually other refinements, like the effect of projectile shape and spin, were built into calculations. However, the most basic result of all—a first-approximation parabolic trajectory— was established by Galileo with his crystal-clear explanation of its dependence on independent horizontal and vertical motions. His use of his theory to probe properties of projectile motion and give numerical tables provided a fine methodological example for other scientists to follow.

4.6 PREDICTING TIDES

Tides have always been of importance for sailors and those managing port facilities. People living near the oceans and seas need to know when their houses are safe from floods and how best to fish, maintain oyster beds, and otherwise interact with water and beaches. Along with this, there has always been a curiosity about the origin of tides and a desire to find a useful understanding of the mechanism driving them. (The book by Cartright is an excellent introduction to this subject. The 1882 paper by Lord Kelvin and the 1953 paper by Doodson tell the story behind this topic as seen by two of the scientists involved.)

The basic mechanism for the tides was set out by Isaac Newton in his *Philosophiae Naturalis Principia Mathematica*: it is the slight variation over the surface of the earth in the gravitational force exerted on water by the moon and the sun, along with the rotation of the earth, that produces the tides. Many people followed Newton in developing the mathematics of tides, with Laplace making outstanding advances. The problem is difficult and made almost impossibly complex by the great variations in topography limiting the motion of oceans and rivers. Only the very simplest of situations can be analyzed from first principles in any detail. However, the theory does supply crucial data to be used in methods for predicting water levels generated by tides.

4.6.1 A Simpler, Pragmatic Approach

Suppose we ask for the water level at some particular place; the port of Dover, for example. We can use a tide gauge to make records of the level over a long time period and, although such records will reveal intricate variations, we may hope to use them to predict levels at some future time. But how to do that? The idea of using harmonic analysis was introduced by Sir William Thomson, later Lord Kelvin. He was a great believer in the general methods set out by Fourier (which I will discuss in detail in chapter 12), so naturally he suggested that at time t the water level or height $H(t)$ should be represented by a sum of components each of which varies harmonically with a particular frequency. Mathematically, assuming ten components are required, we write $H(t)$ in terms of sine functions as

$$H(t) = A_1 \sin(\omega_1 t + \theta_1) + A_2 \sin(\omega_2 t + \theta_2) + A_3 \sin(\omega_3 t + \theta_3) + \dots + A_{10} \sin(\omega_{10} t + \theta_{10}). \quad (4.3)$$

The nth component contributes an amount oscillating in time with angular frequency ω_n (and hence period $2\pi/\omega_n$). The strength of its contribution is measured by its amplitude A_n. The phase angle θ_n specifies how it is shifted in time. If the set of frequencies is given, then the amplitudes and phases are to be chosen so that the total sum of components matches the tidal record for a particular site. Two questions obviously arise: How do we choose the frequencies? How do we find the appropriate amplitudes and phases?

It may not be possible to predict tides using the complete gravitational theory, but knowing that the moon and the sun are the drivers of the tides tells us that the frequencies ω_n, involved in their various orbital motions, are the ones to use in equation (4.3). There are also shallow-water motions that introduce nonlinear effects and double frequency terms. In Thomson's (Kelvin's) language (see appendix B in his book coauthored with Tait) the constituents in the theory are:

1. The mean lunar semi-diurnal.
2. The mean solar semi-diurnal.

3. The larger elliptic semi-diurnal.
4. The luni-solar diurnal declinational.
5. The lunar diurnal declinational.
6. The luni-solar semi- diurnal declinational.
7. The smaller elliptic semi-diurnal.
8. The solar diurnal declinational.
9. The lunar quarter-diurnal, or first shallow-water tide of mean lunar semi-diurnal.
10. The luni-solar quarter-diurnal, the shallow-water tide.[20]

The next step is to find the amplitudes and phases using the data in the tidal records. Sir William Thomson proposed using the method of harmonic analysis, which allows a general function to be decomposed into a number of periodic components as explained above. (This is a technical matter, and the interested reader is referred to Cartright (chapter 8); Thomson's 1882 lecture; Thomson and Tait (articles 57 to 77 and appendix B, part 7); the expository article by Tony Phillips; and the discussion of Fourier methods that we will come to later in chapter 11. A very instructive example of the tides at San Diego is given on the website overseen by the Center for Operational Oceanographic Products and Services.)

Thus a method is established in which the astronomical data and tidal records are used to give a formula for the tidal heights as in equation (4.3). We must now recall that the 1870s are well before our electronic computer era and realize what a difficult and tiresome business it is to calculate $H(t)$ using that equation.

4.6.2 Tide-Predicting Machines

Sir William Thomson lived at a time when machines were being developed for carrying out calculations, and he drew together various ideas to design a tide-predicting machine. Thomson's machine is shown in figure 4.7. Essentially a rope moves up and down, controlling a pen whose trace indicates the required height $H(t)$. The rope threads around a series of pulleys, each of which mechanically represents one of the components in the tide-predicting formula (so ten pulleys for equation (4.3)). The pulleys move up and down in a harmonic motion at the defined frequencies, and the motion is set to the formula amplitudes and phases. Finally, in the complete machine, the harmonic-motion-generating devices are all linked by a series of gear wheels to a shaft, and turning the shaft is equivalent to time evolution in the tide-predicting formula. (For sketches of the complete mechanism see figure 11.1 in the Smith and Wise biography of Kelvin; Thomson's 1882 lecture; and the *Wikipedia* article that is cited. Photographs of actual machines are in the *Wikipedia* article and in the Parker paper.) Today we call Thomson's machine an analogue computer.

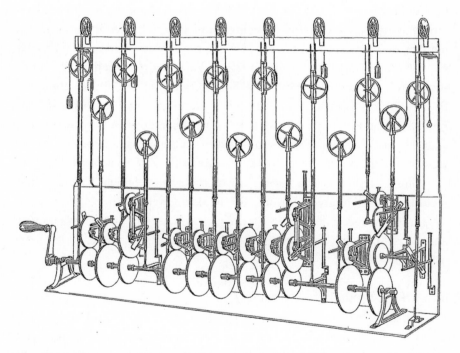

Figure 4.7. Thomson's tide-predicting machine. *From* Wikipedia, *user Terry0051.*

Tide-predicting machines stand as a gleaming brass tribute to Thomson's (and others) ingenuity. The ten-component machine was built with the help of Edward Roberts and Alexander Légé in 1872. Thomson's fifteen-component machine could run off a year's worth of data in about twenty-five minutes. Other machines were built, for example, William Ferrel, in the United States, constructed a nineteen-component predictor in 1882. Later machines built in the twentieth century used up to forty components. The tide-predicting machines were widely used in several countries and, once local conditions were matched, they produced accurate data.

4.6.3 A Wartime Challenge

Knowledge of tides played a crucial part in military planning in the Second World War. A turning point came in 1944 when the Allies planned to invade France. Hitler ordered Field Marshal Rommel to prepare defenses against such an invasion, and Rommel responded by placing thousands of obstacles of all types on the beaches likely to be used. (The paper by Bruce Parker is a wonderful account of this part of history.) Those planning the invasion thus needed detailed information about all the conditions facing an invading force. The magnitude of the task is explained by Parker:

> The Allies would certainly have liked to land at high tide, as Rommel expected, so their troops would have less beach to cross under fire. But the underwater obstacles changed that. The Allied planners now decided that initial landings must be soon after low tide so that demolition teams could blow up enough obstacles to open a corridor through which the following landing craft could navigate to the beach. The tide also had to be rising, because the landing craft had to unload troops and then depart without danger of being stranded by the receding tide.
>
> There were also nontidal constraints. For secrecy, Allied forces had to cross the English Channel in darkness. But naval artillery need about an hour of daylight to bombard the coast before landings. Therefore, low tide had to coincide with first light, with the landings to begin one hour after. Airborne drops had to take place the night before, because the paratroopers had to land in darkness. But they also needed to see their targets, so there had to be a late-rising Moon.[21]

These constraints had to be built into one of the most important calculations ever made. The range of data made available for one of the Normandy beaches is shown in figure 4.8.

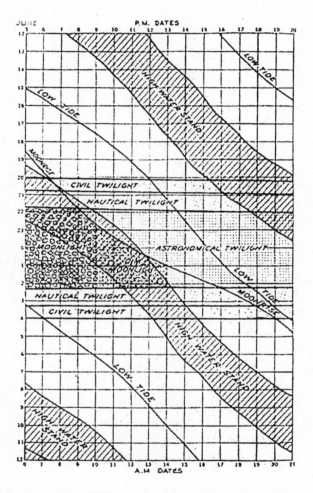

Figure 4.8. Data on tides and light conditions for Omaha beach, June 5–21, 1944. Parker's caption: "Tidal and illumination diagram for Omaha beach, 5–21 June, 1944, shows one of the formats in which Doodson's predictions were provided to military commanders. The diagram gives not only tides but also moonlight and degrees of twilight. Times are given in Greenwich Mean Time." *Reprinted with permission, from "The Tide Predictions for D-Day," by Bruce Parker,* Physics Today *(September 2011). © 2011, American Institute of Physics.*

The tight requirements meant that D-Day could only be on the 5th, 6th, or 7th of June, 1944. (In fact weather conditions led to the 6th being chosen.) The crucial tide calculations were in the hands of Arthur Doodson, one of the great figures in this area of research. He used Thomson's 1872 machine (overhauled in 1942 to handle twenty-six components) and a Robert-designed machine built in 1906, which incorporated forty components. It is hard to imagine that critical decisions made in World War II were linked directly to the initiatives of Lord Kelvin some seventy years earlier. Surely no one will quibble with my choice of **calculation 12, tide predictions**.

4.7 OTHER CANDIDATES

Some readers may call for other calculations affecting life on Earth to be included. An obvious example is weather forecasting, which has gradually become more reliable as fast electronic computers became available. Like most other phenomena on Earth, weather forecasting is extremely complex and driven by available input data.

The book by Wainwright and Mulligan gives a careful introduction to environmental modeling, and a set of contributory papers provides many examples. The theme of the book is "finding simplicity in complexity." This is an area where great progress is being made, but the approximations used to give viable models will always be the subject of debate. Surely Perry and Lord Kelvin would approve. At the forefront of such activities is the question of climate change and the prediction of future temperature rises and such things as the extent of the polar ice cover. There will continue to be intense scrutiny of the results of various climate models, but one day I am sure they will make the lists of great calculations.

4.8 STYLES OF CALCULATION

This has been a long chapter, but before moving on, it is useful to look back and note the very different ways that calculations are made. We have seen calculations:

- that involved only simple arithmetic;
- that required the solution of a differential equation and the use of assumptions and observed data to fit the solution to the problem being investigated;
- that needed geometrical ray tracing, evaluation of times along ray paths, and the matching to experimental data;
- that used diagrams to represent physical processes and required the use of geometry to analyze them;
- and that made use of known data to suggest a formula which could be evaluated using an analogue computer.

Chapter 5

THE SOLAR SYSTEM:
THE FIRST MATHEMATICAL MODELS

in which we see the roles played by mathematical models in describing and understanding the solar system; and meet some of the great scientists involved.

In this chapter, I turn to questions about our home, our planet Earth, in the larger framework of its position in the solar system. Many people now live in cities and regions where light pollution hides the dramatic nature of the night sky. Anyone camping out in remote locations experiences the breathtaking beauty and grandeur of the night sky, which must have equally enthralled, and perhaps overawed, our ancient ancestors. It is also the case that today we do not rely so much on the sky for navigation and for ideas about what is happening to us and why that might be so (although astrology columns still appear in many newspapers and magazines). Nevertheless, we are still fascinated by the stars and the planets, and things like the missions to Mars still make big news.

Thus, in ancient times, the elements of the night sky were more familiar and a cause of wonderment and curiosity. In particular, people were familiar with the objects in the solar system: the sun, the moon and the five visible planets and their daily, monthly, yearly, or longer-time motions viewed with the fixed stars as a background. It is no surprise then that astronomy, both observational and theoretical, played a major part in ancient science. The calculations in this chapter trace the evolution of that ancient astronomy into our modern picture of the solar system.

5.1 AN EARLY PINNACLE

All ancient civilizations were interested in astronomy and astrology. Over the centuries, a large number of observations were collected, and astronomical ideas and calculations were accumulated. (See Thurston or Pedersen for a concise summary.) This culminated with the publication around 150 CE of Ptolemy's *Almagest*, which reviews and builds on early work. Ptolemy's *Almagest* may be thought of as the astronomical equivalent of the mathematical compilation forming Euclid's *Elements*. Strangely, we know little of the personal lives of either of those great writers. Claudius Ptolemy almost certainly lived around the years 100 to 175 CE in Alexandria in Greco-Roman Egypt. He was a brilliant and highly productive man with published works ranging over astronomy, optics, musical theory, astrology, geography, and cartography.

The original Greek title of Ptolemy's book was *Mathematical Compilation*, and later it became known as *The Great* (or *Greatest*) *Compilation*. Like so much early science and mathematics, it was preserved in the Arab world with the title *Al-majisti*, and later in the medieval translation into Latin it became *Almagestum*, and hence today we use *Almagest*. To see why it is simply "the greatest," we need only look at G. J. Toomer's magnificent translation, a book running to over six hundred pages. (Toomer's biographical article is also a standard reference.) In Toomer's words, the *Almagest*

> is a manual covering the whole of mathematical astronomy as the ancients conceived it. Ptolemy assumes in the reader a knowledge of nothing beyond Euclidean geometry and an understanding of common astronomical terms; starting from first principles, he guides him through the prerequisite cosmological and mathematical apparatus to an exposition of the theory of the motion of those heavenly bodies which the ancients knew (Sun, Moon, Mercury, Venus, Mars, Jupiter, Saturn, and the fixed stars, the latter being considered to lie on a single sphere concentric with the earth) and of various phenomena associated with them, such as eclipses.[1]

5.1.1 Ptolemy's Starting Points

Ptolemy has a general discussion in book 1 (the *Almagest* is divided into thirteen books) in which he sets the scene and covers some basic assumptions. Some section headings will illustrate his thinking:

- That the heavens move like a sphere.
- That the earth too, taken as a whole, is sensibly spherical.
- That the earth is in the middle of the heavens.
- That the earth does not have any motion from place to place, either.

Ptolemy is following Aristotle and others in using a geocentric approach with the circle as the definitive geometrical element. Since the heavens must be perfect, it was long argued that motions there must also be perfect—which, to the ancient Greeks, meant uniform motion in a circle.

Thus the mathematics that Ptolemy needed is that relating to circles and spheres. In book 1, he gives the essential details, and in book 2, he tells us how to use them. The geometry of Euclid gives properties of circles, but for quantitative work, Ptolemy needs what today we call trigonometry, which allows angles and lengths to be calculated and manipulated. Ptolemy is recognized as one of the founders of trigonometry.

The basic element in Ptolemy's trigonometry is the chord of an angle, as shown in figure 5.1. If two radii in a circle form an angle α at the center, the chord of that angle is the length of line AB where A and B are the points at which those radii cut the circle. Ptolemy gives results assuming a circle of radius 60, so in terms of the modern sine function, for an angle α,

$$chord(\alpha) \; = \; ch(\alpha) \; = \; 120 \sin(\tfrac{1}{2}\alpha). \qquad (5.1)$$

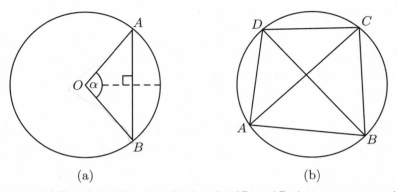

(a)   (b)

Figure 5.1. (a) The chord of angle *a* is given by *AB*, and Ptolemy assumes a circle with radius 60. (b) The quadrilateral *ABCD* as used in Ptolemy's theorem. *Figure created by Annabelle Boag.*

Other Greek mathematicians had calculated chords, but in the *Almagest*, we find the first truly comprehensive table giving all chords for angles ranging up to 180° and going in steps of ½°. Here are the first few entries, taken from the table in Toomer's translation:[2]

Arcs	Chords	Sixtieths
½	0 31 25	1 2 50
1	1 2 50	1 2 50
1½	1 34 15	1 2 50
2	2 5 40	1 2 50
2½	2 37 4	1 2 48

The first column gives the angle in degrees, and the second column gives its chord written in terms of the base 60 system introduced by the Babylonians as we saw in chapter 2. Thus

$$ch(2) = 2 \ 5 \ 40 \equiv 2 + \frac{5}{60} + \frac{40}{60^2} = 2.09444.$$

(Now we can appreciate why Ptolemy used a circle with radius 60 in his definition of chords.)

The third column, also in base 60, tells us just how thorough and detailed Ptolemy's results are. The entries $s(\theta)$ are defined by

$$s(\theta) = \left(\tfrac{1}{30}\right)[ch(\theta + \tfrac{1}{2}°) - ch(\theta)].$$

So $s(\theta)$ is the average increase of the chord over the half-degree interval divided by 30, and hence corresponds to the increase for one minute of arc. The values of $s(\theta)$ may be used in an interpolation calculation so that actually all angles up to 180° are effectively covered in one- minute intervals. What a brilliant achievement!

Ptolemy's definitive table would be used for centuries and by now you are probably beginning to see why he features in this book. For the interested reader, a few calculation details are given in the next section, and the books by Van Brummelen and Chabert should be consulted for further discussions.

5.1.2 Calculating Chords

I now briefly summarize how chords may be calculated according to Ptolemy. First, we note that Euclid gives results about squares, pentagons, and hexagons inscribed in circles, and using such results leads to values of the chords for angles 36°, 60°, 72°, 90°, 120°, and 180°.

Ptolemy gave his own theorem in Euclidean geometry: for the quadrilateral as shown in figure 5.1 (b), the sides and diagonals are related by

$$DB \times AC = AB \times DC + AD \times BC.$$

Using this result with correctly chosen quadrilaterals leads to results for chords of the difference of two angles, $ch(\alpha - \beta)$, and half angles, $ch(\tfrac{1}{2}\alpha)$, that are equivalent to the well-known sine formulas:

$$\sin(x-y) = \sin(x)\cos(y) - \cos(x)\sin(y),$$
$$2\sin^2(x/2) = 1 - \cos(x).$$

Ptolemy could then find, for example, the chord of $72 - 60 = 12°$ and follow that with the chords for $6°$, $3°$, and $1\frac{1}{2}°$. Ptolemy also gives a formula for the chord of the sum of two angles. He came up with a clever scheme using inequalities to find $ch(1°)$ (see Van Brummelen or Chabert), and then using the above results, he could construct the whole table.

5.1.3 Describing the Solar System

Ptolemy's geocentric picture of the solar system is illustrated in figure 5.2. In the *Almagest*, he describes how to calculate the positions of all the celestial bodies in terms of circles in various combinations. He uses the known observational data to find the model parameters, and then he produces tables that may be used to predict future astronomical observations. This is a mammoth task; I defy anyone to look through the *Almagest* and not be stunned by Ptolemy's achievement and the level of detail involved. (The book by Linton gives a good introduction to Ptolemy's theories. On the downside, there are claims that Ptolemy fudged some of his data—see Linton, page 70, for references to the debate on that.)

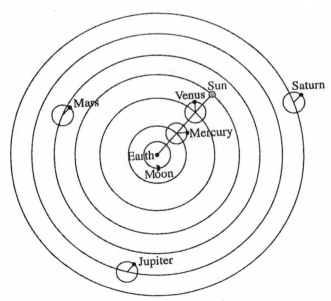

Figure 5.2. Ptolemy's geocentric picture of the solar system. *Reprinted with permission, Cambridge University Press, from C. M. Linton, From Eudoxus to Einstein: A Mathematical History of Astronomy (Cambridge: Cambridge University Press, 2004).*

I will give three examples of Ptolemy's astronomical calculations, beginning with the sun. Recall that the orbits must involve circles and uniform motions. Now, the sun does not move uniformly on a circle centered on the earth. To correct for that, Ptolemy displaces the circle center to a point O as shown in figure 5.3. The sun moves uniformly on that circle so the angle θ is a linear function of time. The sun's angle as viewed from the earth, θ_E in figure 5.2, does not vary uniformly in time, and hence the observed variable speeds of the sun in its orbit are modeled. Ptolemy must show how the "prosthaphaeresis" angle δ can be calculated so the observation angle $\theta_E = \theta - \delta$ is known. (Van Brummelen gives a detailed explanation of Ptolemy's working in chapter 2 of his book.) With the model established, Ptolemy can produce tables for the position of the sun.

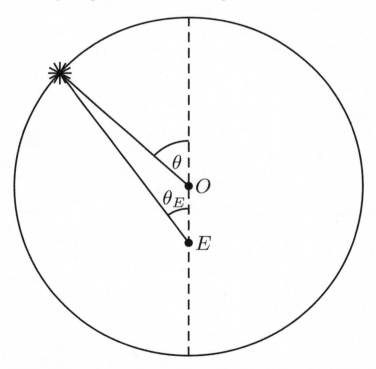

Figure 5.3. Path of the sun as observed from the earth E. The path is a circle centered on O, and the sun moves uniformly around that circle. The angle θ increases linearly in time, but the angle θ_E observed from the earth does not. *Figure created by Annabelle Boag.*

My second example is Ptolemy's treatment of planetary motion. The motion of the planets (as observed from the earth) is even more complicated, and retrograde motion can occur in which the planet appears to reverse direction in its path seen against the background of the fixed stars. To handle this, Ptolemy uses the epicycle device as shown in figure 5.4.

Ptolemy stays true to the old guiding principles—uniform motion on circles—but now a more involved combination of such motions must be used. The situation for a superior planet (Mars, Jupiter, or Saturn) is shown in figure 5.4. The planet P moves uniformly around a small circle (the "epicycle"), whose center C moves in the same sense around a larger circle (the "deferent") with center O. The center O is displaced from the earth E. There is a new point (the "equant") Q, and it is the radial line QC that rotates uniformly, not the line OC joining the two centers. By fitting the model parameters to observed data, the properties of the planet's motion can be matched, and the motion around the epicycle accounts for the observed retrograde motion.

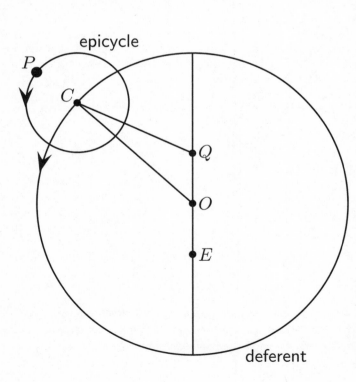

epicycle

P

C

Q

O

E

deferent

Figure 5.4. Orbit scheme for a superior planet (one whose orbit lies outside Earth's). The deferent has center O, and the earth E and the equant Q are equally but oppositely displaced from O. The planet P moves uniformly around the epicycle, whose center C moves around the deferent. *Figure created by Annabelle Boag.*

My third example is Ptolemy's model for the motion of the moon. The motion of the moon is complicated because both the sun and the earth have a strong controlling influence over it, as Isaac Newton discovered. In fact, those complications caused Newton such trouble that allegedly he told mathematician John Machin that "my head never ached but with my studies of the moon," and he told Halley that the theory of the moon "has broke my rest so often I will think of it no more."[3] Ptolemy's mathematical model for the motion of the moon is an epicycle model like that in figure 5.4, but now: the moon travels around the epicycle in a direction opposite to that of the center C around the deferent; the center O of the deferent itself moves around the earth on a new circle. (See figure 3.12 in Linton.) This intricate model is a tribute to his inventiveness, and fitting the various parameters would have been a most complex business. It seems clear that despite Ptolemy's first intentions, this is now well beyond the simple "uniform motion around a circle" dictate of ancient Greek philosophy, and it is remarkable that he could devise such a model to operate so well and to use in such things as eclipse calculations.

One final point: the idea of describing planetary motion in terms of a curve centered on the earth and constructed using epicycles may become clumsy, but it is mathematically sensible. An interesting commentary is given by Hanson in "The Mathematical Power of Epicycle Astronomy."

5.1.4 An Unparalleled Achievement

As mentioned earlier, even a brief examination of the *Almagest* may leave the reader awestruck—and quite rightly so. It is an achievement without equal in ancient science. Of course, there were amendments and refinements made by Arab and other astronomers, but basically the *Almagest* was the guide to astronomy for the next 1,400 years. The first printed version appeared in Venice in 1515 CE. The most useful tables in the *Almagest* were edited and published separately in the *Handy Tables*. The *Handy Tables* were translated into many languages and had a larger circulation than the *Almagest* itself; in historical terms "their longevity, wide distribution and influence among astronomers worldwide mean that Ptolemy's *Handy Tables* can justifiably claim to be the first mass-produced mathematical table."[4]

Surely no one could doubt that the list of candidates for the title of the great calculations must include **calculation 13, Ptolemy's *Almagest*.**

5.2 THE GREAT STEPS

The theoretical astronomy discussed above was based on two key ideas: the earth is the center of the solar system (geocentric model); and the mathematics to be used must involve uniform motion in circular paths. It is clear that Ptolemy was struggling to satisfy those requirements, and ingenious devices were devised to build them into his mathematical models. Astronomy could make giant steps forward only when those key ideas were challenged and overthrown.

5.2.1 The Sun Becomes Supreme

There were ancient Greeks, like Aristarchus, and later mathematicians, for example Nicole Oresme (1320–1382), who suggested that the sun, rather than the earth, should be at the center of the solar system and that the earth is in motion. However, it is in some ways easier to match observations made on Earth using a mathematical model centered on the earth, and we have seen that Ptolemy did just that with considerable success. But there are problems with a geocentric model, and after many centuries it was becoming evident that a more accurate description of the solar system was needed. The great shift to a heliocentric viewpoint was made by the Polish astronomer Nicholas Copernicus (1473–1543) in his *On the Revolutions of the Heavenly Spheres*, the published form of which he saw on his deathbed. The Earth now rotated and moved on an orbit like the other planets.

A model with both the earth and the planets orbiting the sun gives a simple explanation of a planet's retrograde motion (see figure 5.1 in Linton). It also explains why we see phases for Venus (famously observed by Galileo) as shown in figure 5.5. The hope is that the mathematical details of the orbits will also be simpler so that more accurate tables may be produced. But Copernicus still held to the ancient Greek mandate that motion in the heavens must be perfect and involve only uniform motion in circles. The calculations of orbits by Copernicus still used devices like epicycles following the same pattern as Ptolemy. Linton suggests that "Copernicus might well be described as the last of the ancients, a spiritual companion of Aristarchus, Hipparchus and Ptolemy."[5]

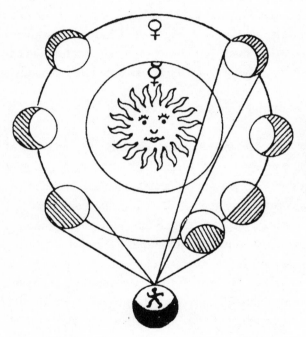

Figure 5.5. Why we see phases for Venus. *From Johannes Kepler,* Epitome Astronomiae Copernicanae *(1621).*

Copernicus knew that he lacked direct proof that the earth was indeed in motion, and he was aware that without it, he would encounter the hostility of the Catholic Church to any suggestion that the earth should lose its preeminent role as the center of the universe. The Lutheran theologian Andreas Osiander took care of the printing of Copernicus's book, and without the author's knowledge, he inserted these highly significant words into a preface:

For these hypotheses need not be true or even probable. On the contrary, if they provide a calculus consistent with observations, that alone is enough. . . . For this art, it is quite clear, is completely and absolutely ignorant of the causes of the apparent nonuniform motions. And if any causes are devised by the imagination, as indeed very many are, they are not put forward to convince anyone that they are true, but merely to provide a reliable basis for computation.[6]

Copernicus had updated Ptolemy's approach by changing from an Earth- to a sun-centered solar system, and Osiander is asking readers to forgive him that step because he is only trying to construct the best mathematical model. (Of course, we do not know Copernicus's reaction to Osiander's intrusion since he was dying when the book came into his hands.)

5.2.2 Enter the Game Changer: Johannes Kepler

Occasionally in history there is someone who changes the direction of science is an important and profound way. Such a man was Johannes Kepler (1571–1630). He was born in Germany and educated at Tübingen, where his university studies covered astronomy, mathematics, astrology, and theology. His intention was to be a clergyman, but through direction and opportunity, he took on a career as mathematician and astronomer. He remained deeply religious and something of a mystic. Kepler traveled widely in Europe and spent periods in Graz and Prague. His eventful and colorful life (at one time he had to defend his mother against charges of witchcraft) is described in the *Dictionary of Scientific Biography* article by Gingerich, who gives a comprehensive bibliography covering everything about Kepler and his work. The proceedings of a conference marking the 400-year anniversary of Kepler's work were edited by Arthur and Peter Beer to create an encyclopedic, thousand-page reference and additional resource for the Kepler addict.

Kepler was an ardent Copernican, but his work was driven by a new approach: he wished to go beyond the mathematical models of the astronomer to appreciate the underlying mechanisms sought by the physicist. In 1596, he published *Mysterium Cosmographicum* with his intentions revealed in the introduction:

And there were three things above all for which I sought the causes why it was this way and not another—the number, the dimensions, and the motions of the orbs.[7]

In summary, Kepler still used mathematical models, but now the search for underlying physical causes was to be used to explain and motivate those models.

In Kepler's time there were the five planets known in ancient times plus the new Copernican planet Earth. The natural questions for Kepler were: Why just six planets? And why are they spaced out as we observe them? These are the mysteries he tackled in *Mysterium Cosmographicum*. In a demonstration of his brilliance and inventiveness, Kepler noted that Euclid concluded the *Elements* by showing that there can be only five regular polyhedrons—or perfect or Platonic solids, as they are sometimes called. (The faces of a regular polyhedron are all identical and must be regular polygons. See figure 5.6.) Furthermore, Kepler found that those five solids could be nested inside one another with a sphere drawn at each interface, see figure 5.6. He then suggested that the planets are arranged on these spheres and calculating the resulting spacing from the geometrical arrangement, he found quite reasonable agreement (within 5 percent) with the accepted planetary distances.

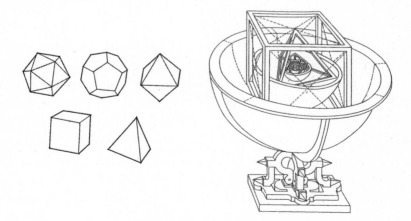

Figure 5.6. The five perfect polyhedrons, and Kepler's system of nested polyhedrons and accompanying spheres. *From Johannes Kepler,* Mysterium Cosmographicum *(1596).*

Kepler's result is stunning; a mathematical classification result in geometry is used to explain why the number of planets is limited and why they are spaced out as in the solar system. Of course, the discovery of more planets spoils Kepler's scheme, but it remains as an impressive and ingenious calculation.

5.2.3 Revolutionary Advances

The Danish astronomer Tycho Brahe (1546–1601) recorded an enormous collection of data and raised the accuracy of observational astronomy to a new level. Kepler met Brahe in 1600 and took over as imperial mathematician when Brahe died. Kepler was charged with accounting for the orbit of Mars and the production of new astronomical tables, the *Rudolphine Tables*, named for the Emperor Rudolph. So began the battle with Mars.

Kepler's work using the data on Mars is one of the epic calculations involving great dedication, supreme technical abilities, and a new and exemplary respect for scientific principles. He began with the usual Copernican orbits with a range of variations as he tried to fit Brahe's superb data. For each new model, Kepler had to use data to fit the model parameters, and then he tested it against other data. The thoroughness of his work and the level of accuracy he demanded are remarkable. One point must be emphasized: Kepler refused to be satisfied with models resulting in small errors that others at that time would readily ignore. Also, unlike most scientists (particularly present-day ones), Kepler documented his failings and tortuous path, although few readers would ever wish to follow it all in detail; in fact, at one point he wrote: "if you are wearied by this tedious procedure take pity on me who carried out at least seventy trials."[8] The story can be read in the references to Gingerich (his *Physics Today* article is beautifully written and illustrated), and also in Linton, Thurston, and in Koyré's major study. It is hard for us today to imagine how one man could produce the enormous number of detailed calculations like those taken from one of Kepler's notebooks for figure 5.7.

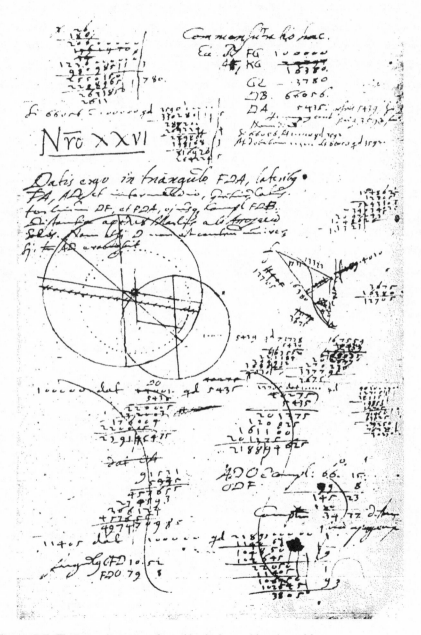

Figure 5.7. The opening page from Kepler's workbook on Mars.
Courtesy of Owen Gingerich.

Kepler's results were published in his 1609 *Astronomia Nova,* known today with its subtitle as *New Astronomy Based upon Causes, or Celestial Physics, Treated by Means of Commentaries on the Motion of the Star Mars, from Observations of Tycho Brahe.* After trying many mathematical forms and thinking about how a magnetic force exerted by the sun might move the planets, Kepler came to chapter 58 where he wrote:

> I was almost driven to madness in considering and calculating this matter. . . .
> With reasoning derived from physical principles agreeing with experience,
> there is no figure left for the orbit of the planet except for a perfect ellipse.[9]

At last, the tyranny of the circle was vanquished. The uniform motion constraint was also replaced with something completely new: the equal areas in equal times law. Since Kepler believed there was a universal underlying mechanism, he could state that what he discovered for Mars would be true for all planets. So it is that we have Kepler's laws (see figure 5.8), in modern form:

Law 1: The orbit of each planet is in the shape of an ellipse with
the Sun at one focus.
Law 2: In any equal time intervals, a line from the planet to the
Sun will sweep out equal areas.

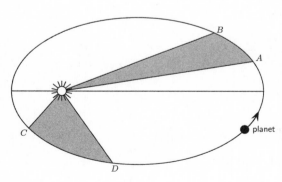

Figure 5.8. The orbit of a planet around the sun according to Kepler's laws. The orbit is an ellipse with the sun at one focus. If the planet moves from *A* to *B* in the same time that it takes to move from *C* to *D*, then the shaded swept-out areas are also equal. (The elliptical shape is greatly exaggerated, and the actual orbits are quite close to circular.) *Figure created by Annabelle Boag.*

Kepler continued to look for the overall harmony of the solar system, sometimes using musical analogies. In 1618, he published *Harmonice Mundi,* in which he reported that he had finally found the rule linking all planetary orbits in what we now call Kepler's third law:

Law 3: The ratio of the squares of times of revolution of any two planets around the Sun is proportional to the ratio of the cubes of their mean distances from the Sun. (Or as sometimes stated, the square of the period is proportional to the cube of the orbit size.)

Using his new theory, Kepler produced the *Rudolphine Tables* in 1627, and astronomers now had a new resource of unprecedented accuracy. (Incidentally, in producing those tables, Kepler was able to use the idea of logarithms introduced in section 3.2.) The equation giving orbit position for a specified time is now known as Kepler's equation, and it is notoriously difficult to solve (see Pask, chapter 12); Kepler tackled this problem and included in his tables results that could be used in an interpolation scheme.

5.2.4 The Achievement

It would be hard to overestimate the significance of Kepler's achievements; he changed the course of science with his ideas about physical reasoning, and his insistence upon fitting data with great accuracy set the standard for science ever after. Most people hear of Kepler's laws as a product of Newton's dynamics and theory of gravity; it is too easily forgotten that they were discovered on the basis of vast and truly heroic calculations. I add **calculation 14, Kepler's astronomical calculations** to my list of great calculations.

Kepler set out what has become our modern view of the solar system. The next steps required the discovery of the underlying mechanism and the proof that the theory matched centuries of observations. This takes us into the next chapter.

Chapter 6

THE SOLAR SYSTEM: INTO THE MODERN ERA

in which we see the roles played by six calculations in establishing our modern theory of the solar system; and meet some of the great scientists involved.

The previous chapter ended with Kepler completing the revolution begun by Copernicus and suggesting the next steps that needed to be taken. We now move into the era of Newton and Einstein.

6.1 THE REVOLUTION CONTINUES

Kepler had stressed that astronomy should become a branch of physics by investigating the underlying causes of orbital motion. He had no proper theory of dynamics and struggled to use William Gilbert's ideas about magnetic forces on a large scale. The next step was taken by Isaac Newton. In his 1687 *Philosophiae Naturalis Principia Mathematica* (generally referred to as the *Principia*), Newton developed his theory of dynamics and then showed that if it was applied to the solar system with the force of gravity, it would supply the physical basis that Kepler demanded. Thus Kepler's laws became a consequence of Newton's dynamical theory of the solar system. (See Pask for a detailed introduction.)

Newton discovered that his theory of the solar system was successful if the attractive force of gravity between two bodies with masses m and M is given by

$$F = \frac{GmM}{r^2}, \tag{6.1}$$

where r is the distance between the bodies and G is the gravitational constant. However, the question remained: Was the force acting in the heavens (the force acting between the sun and the planets), the same as the force we know as gravity on Earth (the force that accelerates a falling body)? In the *Principia*, Newton set out his "Rules of Reasoning in Philosophy," which include:

> Rule I. We are to admit no more causes of natural things than such
>   as are both true and sufficient to explain their appearances.
> Rule II. Therefore to the same natural effects we must, as far as
>   possible, assign the same causes.

In his discussion of Rule II, Newton writes: "As to respiration in man as in beast; the descent of stones in Europe and America; the light of our culinary fire and of the sun; the reflection of light in the earth, and in the planets."[1] In other words, while he is wrong about the sun being a fire, he is clearly saying that we should assume the same causes for phenomena whether they occur on Earth or in the heavens. This is a major assumption; in fact, it is the central assumption for the whole of astrophysics. Can Newton give evidence to back up his assumption?

6.1.1 The Moon Test

Newton devised the ingenious Moon Test as a way of comparing how bodies fall in different places. This would then allow him to suggest properties of the gravitational force in those places.

Figure 6.1 shows the moon in its orbit around the earth. Suppose that in one minute, the moon revolves through the angle θ. If there was no force operating, the inertia of the moon would see it move off along a straight line tangent to its orbit. However, this movement is supplemented by the fall through the distance D under the influence of the earth's gravitational attraction. Newton finds this fall, D, in one minute, to be $15^{1}/_{12}$ Paris feet. (The details are in section 6.1.2.)

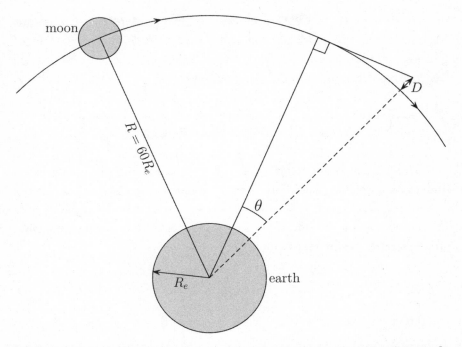

Figure 6.1. A diagram showing the moon traveling through an angular distance θ and falling through a distance D as calculated in Newton's Moon Test. The earth has radius R_e, and the radius R of the moon's orbit is approximately 60 R_e. *Figure created by Annabelle Boag.*

(Incidentally, Kepler, and others, did not have that concept of inertia, which requires a force directed toward the earth to keep the moon in its orbit; Kepler wished to find a force that would push or sweep the moon around its orbit, which would imply a force tangent to the orbit rather than perpendicular to it.)

Newton next considers a body (perhaps an apple!) falling on the surface of the earth. Assume that the same gravitational force (see equation (6.1) operates on the earth and on the moon's orbit. Then, since the moon's orbit has a radius about sixty times the radius of the earth, the inverse dependence on distance squared in equation (6.1) tells us that the force of gravity is sixty squared times bigger on Earth than it is on the moon's orbit. Thus it follows that while the time taken to fall $15^{1}/_{12}$ Paris feet is one minute on the moon's orbit, it should be only one second for a body on Earth. (See below for details.)

Newton notes that measurements by Huygens reveal that in one second, a body on Earth does indeed fall $15^{1}/_{12}$ Paris feet. The force of gravity, equation (6.1), must be the same for the moon and for objects on Earth. Thus Newton triumphantly concludes:

> therefore (by Rule I and II) the force by which the moon is retained in its orbit is the very same force which we commonly call gravity.

This is one on the major steps in the *Principia*, and from it, Newton moves on to declare that he has found the universal theory of gravitation, one of science's greatest discoveries. He also established that the physics we use here on Earth is the physics to use in the heavens. I add **calculation 15, Newton's Moon Test** to my list.

6.1.2 A Few Details

If you wish to check Newton's work, you must begin with his data:

Moon's orbit: radius R is approximately sixty times the Earth's radius R_e; period is 27 days, 7 hours, 43 minutes = 39,343 minutes; angle θ revolved through in one minute is $2\pi/39{,}343$ radians or $(360/39{,}343)°$.

Earth: circumference, $2\pi R_e$ is 123,249,600 Paris feet so R_e is 123,249,600/2π Paris feet.[2]

The triangle containing the angle θ (see figure 6.1) gives

$$\frac{R+D}{R} = \sec(\theta) \quad \text{or} \quad 1 + \frac{D}{R} = \sec(\theta) \simeq 1 + \frac{\theta^2}{2} \quad \text{for small } \theta.$$

So we get $D = R\theta^2/2$ and using Newton's data this comes out as $15^{1}/_{12}$ Paris feet.

The fall of a body on Earth in time t is given by the standard formula $\frac{1}{2}gt^2$. Since gravity g is 60^2 times greater than it is on the moon's orbit, to give the same distance of fall we must make t^2 60^2 times smaller, so t is to be 60 times smaller. Thus a fall taking one minute on the moon's orbit should be compared with a fall on Earth over the period of one sixtieth of a minute, or one second. Hence Newton concludes that if gravity is the same and follows equation (6.1), in both of these cases the fall distance should be $15^{1}/_{12}$ Paris feet. And it is!

6.2 WEIGHING THE PLANETS

I have omitted several calculations of great originality and value in the *Principia*, but my next one is just too unexpected and brilliant in its simplicity to ignore. It concerns the masses of the planets, including the earth, as I mentioned in section 4.2.

The force of gravity, as detailed in equation (6.1), depends on the masses of the two interacting bodies. It seemed clear that the sun is a large body and its mass m_{sun} is much larger than the masses of the planets orbiting around it. But how much larger? Would the mass of any planet be so large that it would have a significant effect on the orbits of others? Would planetary masses be so large that the center of gravity of the solar system might deviate significantly from the center of the sun? Newton came up with a beautiful way to find the masses of the earth and the large planets Jupiter and Saturn using the fact that they each have satellites, or moons, circling them.

Newton's derivation of Kepler's third law allows us to write it as an equation rather than as a ratio property. (Actually Newton worked mostly in terms of ratios, too, but to explain his approach here, it is simpler to use an equation.) If a body orbiting a large body of mass M has period T and orbit semi-major axis a, Kepler's third law gives

$$\frac{T^2}{a^3} = \frac{(4\pi^2 / G)}{M}.$$
(6.2)

(If the mass of the orbiting body is not insignificant in comparison with M, a correction is needed, but that refinement is not of great importance here.)

Suppose that a planet orbits the sun with period T_p and semi-major axis a_p. Further, suppose that a moon orbits that planet with period T_{mn} and semi-major axis a_{mn}. Kepler's third law, equation (6.2), applies to the sun-planet system (with M taken as m_{sun}) and to the planet-moon system (with M now taken as m_p). The result is the two equations

$$\frac{T_p^{\ 2}}{a_p^{\ 3}} = \frac{(4\pi^2 / G)}{m_{sun}} \quad \text{and} \quad \frac{T_{mn}^{\ 2}}{a_{mn}^{\ 3}} = \frac{(4\pi^2 / G)}{m_p}.$$

Eliminating the unknown factor $(4\pi^2/G)$ between these two equations produces

$$\frac{m_p}{m_{sun}} = \frac{T_p^{\ 2} a_{mn}^{\ 3}}{T_{mn}^{\ 2} a_p^{\ 3}}. \tag{6.3}$$

This remarkable result tells us that if we observe the details of a planet's orbit around the sun, and the orbit details for one of its moons, then we can find the mass of the planet as a fraction of the sun's mass. There are moons around the earth, Jupiter, and Saturn, and substituting orbit data into equation (6.3) gives:

	Sun	Jupiter	Saturn	Earth
mass (*Principia*)	1	1/1,067	1/3,021	1/169,282
mass (modern)	1	1/1,047	1/3,500	1/332,480

The table indicates that Jupiter and Saturn are large planets, and in refined calculations of solar system phenomena, it is necessary to take into account their gravitational effects on other bodies. This was a result of great importance, and we shall see an example in the next section. The earth is around a hundred times smaller again, and so, generally less important as a perturbation. The surprising error for the earth, as noted in the table, is due to Newton's poor value for the solar parallax and a copying error in the working. Nevertheless, Newton did establish the vitally important fact that the earth is relatively small compared with those larger planets. (For further discussion, see Pask, chapter 23, and references therein, especially the 1998 paper by Cohen.)

This is one of those calculations that makes me (and you, too, I hope) smile with delight and wonder just how anyone could have thought of such an elegant and powerful method. As Huygens put it at the time, Newton's feat gave us information about the planets that "hitherto has seemed quite beyond our knowledge."[3] Today, the masses of planets without moons can be found in different ways and from spacecraft fly-by data. I name **calculation 16, Newton's determination of planetary masses**.

6.3 A FIRST CONFIRMING TRIUMPH

The ancient view that beyond the earth there were perfect, unchanging heavens was gradually discredited. Kepler observed and wrote about the supernova that flared so brilliantly in 1604, and Galileo used his telescope to reveal that the moon has mountains and valleys just as the earth does. Over the centuries, there had been many sightings of comets, and they were said to predict the coming of dire events. In Newton's time, it was said that comets foretold the tragedies of the plague eruption in 1665 and the great fire of London the following year:

> In the first place a blazing star or comet appeared for several months before the plague, as there did the year after another, a little before the fire. The old women and the phlegmatic hypochondriac part of the other sex . . . remarked . . . that those two comets passed directly over the city and that so very near the houses that it was plain they imported something peculiar to the city alone.[4]

Newton devoted over fifty pages of the *Principia* to the subject of comets. During a preliminary discussion he makes the following points:

- The comets are higher than the moon, and in the regions of the planets.

- The comets shine by the sun's light, which they reflect.
- I am out in my judgment, if they are not a sort of planets revolving in orbits returning into themselves with a perpetual motion.
- That the comets move in some sort of conic sections, having their foci in the center of the sun, and by radii drawn to the sun describe areas proportional to the times.
- Hence, if comets are revolved in orbits returning into themselves, those orbits will be ellipses; and their periodic times be to the periodic times of the planets as the $\frac{3}{2}th$ power of their principal axes.[5]

In short, the suspicious nature of comets has been removed; they are bodies much like planets, and Newton's theory explains their motions. Newton also recognizes the computational problems involved. The orbits of the comets are enormous (see figure 6.2) and we only have access to a very small part of that orbit. He suggests that, since the conics are very similar in that small region (see figure 6.3), a parabola is most easily fitted, and the orbit can be corrected later. After reaching these conclusions, Newton presents a long discussion on observed comets and fitting orbits to the observational data, noting that "this [is] a problem of very great difficulty."

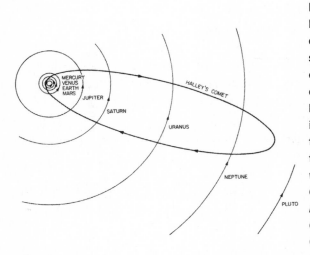

Figure 6.2. The orbit of Halley's Comet projected onto the plane of the solar system. Note the enormous size of the orbit and that the comet has retrograde motion—it travels in its orbit in the opposite direction of the planets. *Reprinted with permission, © Cambridge University Press, from D. W. E. Green,* The Mystery of Comets *(Cambridge: Cambridge University Press, 1986).*

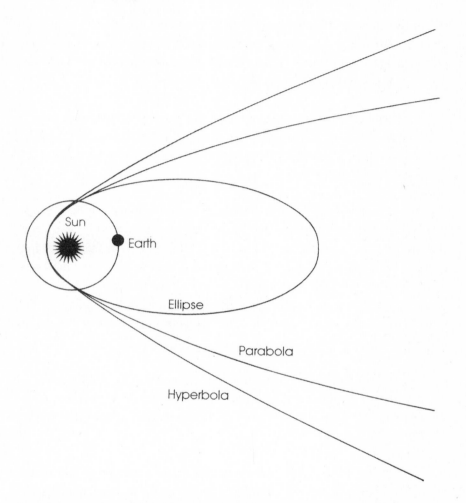

Figure 6.3. An illustration of the similarity of different conic-section orbits near the sun, which is taken as the focus. *Reprinted with permission, © Cambridge University Press, from D. W. E. Green,* The Mystery of Comets *(Cambridge: Cambridge University Press, 1986).*

6.3.1 Enter Edmond Halley

Edmond Halley (1656–1742) was involved in the *Principia* from the very beginning (see Pask chapter 3). Following a meeting with Newton, Halley began the business of extracting from him an account of his work. After much persuading and cajoling, the first edition of the *Principia* was completed in 1687, and it was Halley who paid for its publication. Only one person is acknowledged in Newton's Preface:

> In the publication of this work the most acute and universally learned Mr. Edmond Halley not only assisted me in correcting the errors of the press and preparing the geometrical figures, but it was through his solicitations that it came to be published.

Halley is best known for mathematics and astronomy, but he was also involved in biology, geology, geography, physics, engineering, and deep-sea exploration using his invention of a diving bell. (We shall meet him again in sections 6.5.2 and 8.2. See Ronan for more on this remarkable man.) Newton reported their collaboration in the *Principia*, and the study of the comet that was observed in 1680 shows their theory in action. The fit by Halley to an elliptical orbit is tested in a table comparing calculations and observations. Newton drew this conclusion from the results:

> The observations of this comet from the beginning to the end agree as perfectly as the motion of the comet in the orbit just now described as the motions of the planets do with the theories from whence they are calculated.

Comets had been removed from the realms of superstition and speculation and shown to be as readily described using Newton's dynamics and law of gravity as are the planets.

6.3.2 Halley's Comet

Halley became particularly interested in a comet that appeared in 1682. Certain features of its motion, such as the direction of its path around the sun (see figure 6.2), reminded him of other comets seen in 1531 and 1607. After studying greater details of the orbit, Halley became convinced that only one comet was involved and wrote to Newton that he was "more and more confirmed that we have seen that comet now three times since ye Yeare 1531."[6] Later he found that a comet observed in 1456 also fit the description. Thus it was that he claimed the comet had a period of around 75 or 76 years and estimated that it would appear again in 1758. Halley also suggested that a close passing of Jupiter caused the variability in the period of the comet, and he used 76 years to make his famous prediction.

The calculations for the orbit of this comet became a challenge for the leading astronomers of the time. The most detailed calculations were made by Alexis-Claude Clairaut (see Wilson's appendix to Waff's paper and the book by Grier). Following Newton's work on planetary masses, Clairaut knew that Jupiter and Saturn directly cause perturbations to the comet's orbit, and they also cause variations in the sun's position, which then makes tiny changes in the comet's motion. The perturbations caused by Jupiter and Saturn were extremely difficult to handle, and this was one of the first times that extensive numerical integrations were performed. The extent and the complexity of the problem meant that a new approach to calculations was required, and Clairaut formed a team of calculators with the astronomer Joseph Jérôme Lefrançois de Lalande and Nicole-Reine Étable de Labrière Lepaute, the wife of a noted clockmaker. Clairaut divided the task of moving around the comet's orbit into a series of short steps so that each of them could manage one part of the calculation. For six months, the three of them calculated, often from morning till night. These truly were heroic calculations. Poor Madame Lepaute even missed out on a tribute to her written by Clairaut in his initial report—a jealous lady friend insisted that he remove it.

Clairaut's calculations suggested that the comet would reach its perihelion (closest point to the sun) in mid-April 1759, although there was some uncertainty, perhaps as much as a month, associated with that. As it happened, the comet reached its perihelion on March 13th. Given the complexity of the problem and incomplete knowledge of the solar system at the time, that was surely a remarkable agreement between theory and experiment. Of course there were great rivalries leading people like Jean Baptiste le Rond d'Alembert to belittle Clairaut's work (see the article by Waff), but today we can only wonder at the dedication and persistence shown by Clairaut and his team.

The validity of Newton's theory of universal gravitation was superbly demonstrated with this remarkable meeting of theory and observation. This is how it was announced on April 25, 1759, at the public assembly of the Paris Academy of Sciences:

> The Universe sees this year the most satisfying phenomenon that Astronomy has ever offered us; unique event up to this day, it changes our doubts into certainty, and our hypotheses into demonstrations.[7]

It was at that meeting that the comet was also named Halley's Comet. The appearance of a comet exactly as predicted was surely the most wonderful and convincing evidence that Newton had truly answered Kepler's call and had found the underlying theory for astronomy. The epic struggles involved gave rise to my **calculation 17, predicting the return of Halley's Comet**.

There have been three appearances of Halley's Comet since that wonderful event in 1759; it was carefully tracked in 1835, 1910, and 1986. The calculations for the 1835 return halved the error in Clairaut's great work, and by 1986 the error was measured in hours. (The stories behind all the return calculations are well told by Grier.) My own sense of wonder was still there in 1986 when I too saw a returning Halley's Comet and marveled that a theory could predict a long-term celestial event with such amazing accuracy.

6.4 A SECOND CONFIRMING TRIUMPH

In section 5.2, we saw Kepler struggling to explain why there were just the six planets: Mercury, Venus, Earth, Mars, Jupiter, and Saturn. He was not to know that in 1781 the game would be changed by a musician and amateur astronomer.

William Herschel (1738–1822) was a German musician who moved to England in 1757. Apart from music, his interests included languages, mathematics, and astronomy. He built his own telescopes and established an observatory with the help of his sister Caroline. On March 13, 1781, Herschel observed what he thought was a curious nebulous star or maybe a comet. Future observations showed that its position changed and so it was deemed to be a comet, although one with strange features such as no tail. Observational evidence by Herschel and other astronomers finally led to the opinion that Herschel had stumbled on a seventh planet.

There was some debate about a name for the planet. Herschel favored naming it after King George III, and the French wanted it named Herschel, after its discoverer. Eventually mythology prevailed, and the name Uranus was chosen. (Uranus was the god of the sky, husband of earth, father of Saturn, and so on.)

Even more debate ensued when astronomers began calculating the orbit of the new planet. Gradually a set of observations was established, and a search through the astronomical records revealed that in fact Uranus had been observed by John Flamsteed in 1690, 1712, and 1715; by Pierre Charles Lemonnier in 1750, in 1764, twice in 1768, and six times in 1769; by James Bradley in 1756; and by Tobias Mayer in 1756. Thus several reliable, accurate sightings of Uranus were available, although, of course, none of the observers realized at the time that they were recording the position of a seventh planet.

A problem arose: the tables calculated on the basis of the recent observations did not fit those older ones. In 1820, a former assistant to the great Pierre-Simon Laplace, Alexis Bouvard (1767–1843), produced comprehensive tables and suggested that the older observations had large errors attached to them. However, it was later discovered that the suggested errors were up to ten times larger than those experienced observers were believed to have made. Even worse, within a few years, there were new observations of Uranus that did not fit in with Bouvard's tables.

Astronomers casting around for explanations of Uranus's strange behavior came up with five possibilities. The first three—the Cartesian hypothesis of a cosmic fluid, the existence of a massive satellite around Uranus, and the occurrence of a catastrophic effect like the collision with a comet—were all easily discredited. The fourth possibility was more scientific and interesting: Did the law of gravity (see equation (6.1)) change at the very great distance from the sun where Uranus was located? This possibility had been canvassed before when calculations failed to match observations, but each time it was found that the calculations needed amending or correcting, and Newton's law of gravity reigned supreme.

It was the fifth possibility that seemed to be the answer: maybe there was another massive planet beyond Uranus, and it was perturbing the new planet's orbit so that the standard orbit theory used when constructing tables needed correction. (Clairaut had speculated about very distant planets when he considered effects on Halley's Comet.) Here was one of the great challenges to mathematicians working in astronomy.

(To read more about this episode and its aftermath consult the references to Grosser, Hanson, Linton, Morando, and Smith.)

6.4.1 Aspects of the Calculations

It is worth pausing for a moment to consider the mathematical problems involved. First, there was the question of tables for Uranus; this required finding orbital parameters to fit observations and then computing new data to guide future observations. This was a critical step in revealing the new physical problem—observations just did not fit the conventionally calculated orbit.

If the theoretical tables and observational data were accepted and the discrepancies were due to the influence of an unknown planet, how was that planet to be located? Newton had already explained in his *Principia* that even though planets were much less massive than the sun, their effects in perturbing each other's orbits could be substantial. Thus the branch of applied mathematics known as perturbation theory was to be used (see Linton).

However, for the case of Uranus, there was a twist to the problem: the orbit perturbations for Uranus were not due to some known planet (although those due to Jupiter and Saturn had been considered), but were instead to be used to find the properties of a new planet. Instead of asking what perturbations a particular planet caused, the perturbations were now assumed, and the planet causing them was to be found. This is another classic inverse problem (as discussed in section 4.4.5), with all the difficulties associated with such problems.

Assumptions would need to be made about the plane of the unknown planet's orbit and its mass, as well as about some geometrical details. One way to start was to use the curious Titius-Bode law, which is perhaps better characterized as an empirical formula since it has no real scientific basis. This "law" describes orbit sizes according to

$$\text{distance from the sun} = 4 + 3 \times 2^n.$$

The earth's distance is taken as 10, corresponding to $n = 1$. Venus has $n = 0$, and for Mercury the distance is taken as 4. Mars, Jupiter, Saturn, and Uranus correspond to $n = 2, 4, 5$, and 6, respectively. (The gap at $n = 3$ was filled by the discovery of asteroids.) The predicted values are surprisingly accurate (the error for Uranus is about 2 percent), and so $n = 7$ could be used to suggest a possible orbit size for the unknown perturbing planet.

Finding the origin of the variations in Uranus's orbit was no easy problem, and it became the outstanding mathematical challenge of the time.

6.4.2 Victory to Leverrier

Urbain-Jean-Joseph Leverrier (1811–1877) was a highly talented French mathematician and astronomer who gained experience in perturbation theory early in his career. He published three papers on his work on the orbit of Uranus. In the first two (November 1845 and June 1846), he discussed the perturbations due to Jupiter and Saturn and the remaining discrepancy with Bouvard's tables and made his conclusion:

> I have demonstrated a formal incompatibility between the observations of Uranus and the hypothesis that this planet is subject only to the actions of the Sun and of other planets in accordance with the principle of universal gravitation.[8]

In his third paper (August 1846) Leverrier announced the new, perturbing planet's mass and orbital elements. It was now up to the observational astronomers. There seems to have been little response in France and Leverrier communicated with his friend Johann Gottfried Galle at the Berlin Observatory. Galle, with the assistance of an astronomy student, Heinrich d'Arrest, began the search on September 23rd, the day he received Leverrier's letter. Searching the sky in the area nominated by Leverrier and comparing star charts, it took Galle and d'Arrest about half an hour to locate the eighth planet in the solar system. Other astronomers later confirmed the sighting, and a disc was observed, rather than a twinkling point of light, verifying that the object was indeed a planet.

Here was an enormous triumph for French science (and German astronomers) and another victory for Newton's theory of gravitation. After the usual debate, the planet was given the name Neptune. Leverrier was showered with honors, and, in 1854, he became head of the Paris Observatory.

6.4.3 Another Side to the Story

Leverrier and Galle's discovery of Neptune is one of science's great stories, but there is another series of events that make it also one of the most fascinating and sensational events in the history of science.

John Couch Adams (1819–1892) graduated from Cambridge University as the Senior Wrangler (top student) in the mathematics class of 1843. He had learned about the problems generated by Uranus and the unknown planet hypothesis and resolved to work on a perturbation theory for predicting the nature of the unknown planet as soon as he had graduated from Cambridge and become a fellow of St John's College. Adams was helped by James Challis, the Plumian Professor of Astronomy at Cambridge and head of the university's observatory. After working over the summers of 1843 and 1845, Adams had determined the characteristics required of the new planet and predicted its position.

But now comes the big difference between Adams and Leverrier: Adams did not publish his findings. Knowing Adams's results, Challis corresponded with Astronomer Royal George Biddell Airy, and, at Challis's suggestion, Adams famously went to call on Airy in September only to find him away in France. Adams tried to see Airy again on October 21st, first to find him out and then later in the day to be told Airy could not be disturbed. Adams left a note (Grosser gives the details), but Airy was doubtful about the whole business, and no decisive action was taken. In fact Airy had written that

> With respect to the errors of the tables of Uranus . . . if it be the effect of any unseen body, it will be nearly impossible to ever find out its place.[9]

In short, Airy and Challis procrastinated while Leverrier and Galle gained the glory.

There is a long and complex story involving Adams, Challis, and Airy, which resulted in little action being taken to use Adams's predictions in a search for the unknown planet—particularly until Airy received a letter from Leverrier that showed him Adams was indeed on the right track. Poor Adams apparently prepared a paper to be read at the Southampton British Association for the Advancement of Science in September 1845, but he arrived a day too late for the physical-sciences sessions. The result was that England lost the glory to France, and there was an almighty row in England when all was revealed. Airy was scorned, and the effects lasted until his death in 1890 when thoughts of burial in Westminster Abbey were overturned by memories of his part in the Neptune debacle.

There was a further aspect to this story. Eventually Adams's work was made known, and it was suggested that he should share the credit with Leverrier. It is not hard to imagine the uproar and the vicious exchanges in the now-declining relationship between Britain and France. Figure 6.4 is a French cartoon showing how Adams discovered Neptune by spying on Leverrier across the English Channel. (See references, especially Smith and Grosser, for the full story.) Scientists and scientific societies at the highest level became embroiled in the controversy. Eventually the whole matter was resolved amicably; Adams was also honored, and Leverrier and Adams even became friends.

Figure 6.4. French cartoon showing Adams discovering Neptune by spying on Leverrier's work. *From* L'Illustration *(November 7, 1846).*

6.4.4 Planet Neptune

The discovery of a new planet by means of calculations was a triumph for mathematical astronomers and provided yet more evidence for the validity of Newton's universal theory of gravity. An essential addition to my list is **calculation 18, the discovery of Neptune**.

Before leaving this topic, I turn to a question many readers may be asking: Given the difficulty of the calculations, how good were Leverrier's and Adams's results for the orbit of the unknown planet Neptune? The answer is summarized in figure 6.5.

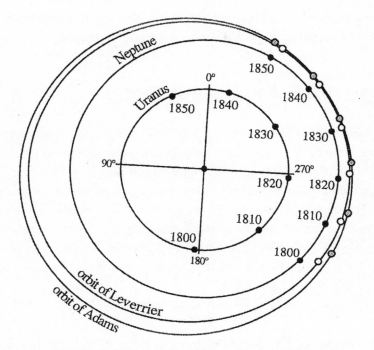

Figure 6.5. The orbits of Uranus and Neptune, and the predictions of Leverrier and Adams. *Reprinted with permission, © Cambridge University Press, from C. M. Linton,* From Eudoxus to Einstein: A Mathematical History of Astronomy *(Cambridge: Cambridge University Press, 2004).*

Clearly the predicted orbits are reasonable (given the technical difficulties) for the years around 1840, and both indicate roughly where Neptune should be found. For the rest of the orbit the results are not too good. However, we must remember that Neptune takes almost 165 years to make one revolution in its orbit, so maybe we should expect the results to be reasonable only in the time period covered in the discovery story.

There is one final twist! The nineteenth century saw the growth of astronomy in America, and, naturally, interest in the discovery of Neptune was as great there as elsewhere. Analysis by Sears Cook Walker, supported by Harvard mathematician Benjamin Peirce, gave rise to the American view that the discovery of Neptune was a "happy accident" and the orbital parameters were quite wrong. Needless to say, there was a new row, this time with America confronting a united Europe. To delve further I recommend the fascinating paper by Kent and the book by Grosser.

6.5 THE SIZE OF THE SOLAR SYSTEM

Over the centuries, accurate measurements were made of the position of the moon and of the planets and the periods of their motions. The fixed stars provide a background for precise angular measurements. However, determining the absolute size of the solar system proved to be extremely difficult, and a great variety of mostly quite inaccurate values were obtained (see Van Helden's book for a detailed historical account). A major step forward came with the discovery of Kepler's third law (see the previous chapter) which tells us that a planet's period T and orbit size a satisfy

$$\frac{T^2}{a^3} = \text{a constant.} \tag{6.2 (a)}$$

If we measure T in earth years and a in terms of the Earth-sun distance (the astronomical unit, AU), then equation (6.2 (a)) for the Earth tells us that the constant is one. Thus for any other planet

$$a = T^{2/3} \text{ AU} \qquad \text{if } T \text{ is measured in Earth years.} \tag{6.2 (b)}$$

For example, the period of Venus is 0.614 Earth years so it has an orbit size of $0.614^{2/3} = 0.72$AU.

The relative size of the solar system is now established, but the absolute size is still unknown. How do we find the value of an astronomical unit, AU, the Earth-sun distance?

Many people, including the eminent astronomers Giovanni Cassini and John Flamsteed, relied on a measurement of the parallax of Mars, but there are difficulties in that method, and the result for the AU was not reliable.

In the second century BCE, Hipparchus had used data from a solar eclipse to find a reasonable value for the Earth-moon distance in terms of the earth's radius (see Linton and Van Helden), but his value for the Earth-sun distance was quite inaccurate. An eclipse of the sun occurs when the moon passes between the earth and the sun, and after Copernicus's change to a heliocentric view of the solar system, it was apparent that two planets, the interior planets, also intervene between the earth and the sun. Could that be used to measure the AU?

(An aside on notation: astronomers really wished to measure the angle subtended at the sun by the earth's diameter, and half that angle is called the solar parallax. Call it β. Since β is small, to a good approximation, the Earth-sun distance, one AU, is given by R_e/β if β is measured in radians, or $180R_e/\pi\beta$ if β is measured in degrees. The modern value of β is 8.797". Note that this is a small angle and thus presents measurement difficulties.)

6.5.1 Transits of Mercury and Venus

When Mercury or Venus passes between the earth and the sun, the result is not an eclipse of the sun but the passage of a small dark spot across its face. This is termed a transit. Transits of Mercury were observed (for example, by Edmond Halley in 1677, acting on the suggestion of James Gregory), but because Mercury is small and close to the sun, it turns out not to be so useful for precise astronomical measurements. That leaves Venus for consideration.

The orbit of Venus does not lie in the ecliptic plane (where we find the sun and the earth), but rather cuts the plane at the ascending and descending nodes. Examination of the orbits in figure 6.6 shows why the transit of Venus is quite a rare phenomenon. For a transit to occur, the earth, Venus, and the sun must line up: Venus must be at the ascending node N_a and the earth at E_1 to give the line E_1N_aS; or Venus must be at the descending node N_d with the earth at E_2 to give the line E_2N_dS. The last eight transits of Venus occurred in 2012, 2004, 1882, 1874, 1769, 1761, 1639, and 1631. Transits occur in pairs separated by about eight years and pairs occur about every 120 years. Kepler famously predicted the 1631 transit, but it was not observed; using Kepler's work, Jeremiah Horrocks predicted and observed the next transit in 1639.

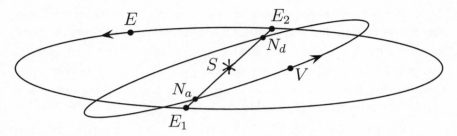

Figure 6.6. Orbits of Earth E and Venus V around the sun S. The orbit of Venus cuts the ecliptic plane at the ascending and descending nodes, N_a and N^d. (Orbit shapes are greatly exaggerated.) *Figure created by Annabelle Boag.*

6.5.2 Halley Shows How to Proceed

Halley was an assistant to Astronomer Royal John Flamsteed for parallax measurements on Mars, and he concluded that the parallax method for finding the astronomical unit was basically useless. Subsequently, in 1678, he wrote:

> There remains but one observation by which one can resolve the problem of the distance of the Sun from the Earth, and that advantage is reserved for the astronomers of the following century, to wit, when Venus will pass across the disc of the Sun, which will occur only in the year 1761 on May 26.[10]

Halley gave a full discussion of the proposed observations and calculations in his 1716 *Philosophical Transactions* paper (see bibliography). The essentials of Halley's method are given with reference to figure 6.7. The transit is to be observed by astronomers at points A and B on the earth where spots at M and N on the sun will be observed. Very large distances are involved, and the angles AVB and MVN in figure 6.7 (a) are equal, leading to

$$\frac{MN}{AB} = \frac{VN}{VA} \quad \text{or} \quad MN = \frac{VN}{VA} \, AB. \tag{6.3}$$

We can use the Venus orbit radius R_v for VN and the difference in Earth and Venus radii $(R_e - R_v)$ for VA. As mentioned above, Kepler's third law tells us that $R_v = 0.72\ R_e$. Using those ideas in equation (6.3) tells us that

$$ MN = \left(\frac{R_v}{R_e - R_v}\right) AB = \left(\frac{0.72 R_e}{R_e - 0.72 R_e}\right) AB = 2.57\, AB. $$

We can find AB (by surveys of the earth), so MN will be known.

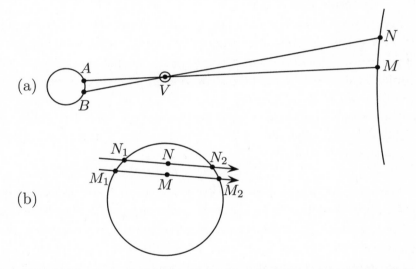

Figure 6.7. (a) Observers A and B on Earth see a shadow spot cast by Venus V on the sun at points M and N. (b) The spots trace out the lines M_1M_2 and N_1N_2 as Venus moves in front of the sun. *Figure created by Annabelle Boag.*

Next we find the diameter of the sun d_{sun} from MN. Now we are almost there: the angular size of the sun θ_{sun} is known by observation, and d_{sun} is simply θ_{sun} multiplied by the AU, the Earth-sun distance. Thus knowing θ_{sun}, and now having found d_{sun}, at last we can find the astronomical unit by calculating d_{sun}/θ_{sun}.

However, there is one problem in all of that: Exactly how do we "find the diameter of the sun d_{sun} from MN"? The angle AVB in figure 6.7 (a) is very small, and it is extremely hard to measure things accurately and carry out the necessary calculations. The whole method appears to be doomed.

Halley made the brilliant suggestion that the two observers should measure not the angles, but rather the time it took for Venus's shadow spot to travel across the sun. Observer A would time the spot moving from N_1 to N_2, and observer B would time the spot moving from M_1 to M_2. See figure 6.7 (b). The angular drift rate for Venus's shadow spot to move across the sun is then known. From these times, the arc lengths can be found and then MN related to the full size of the sun, as required. (See the article by Phillips for the details.)

Halley had shown that the Earth-sun distance could be found by the relatively simple process of measuring two times. Furthermore, he knew from his earlier experience with Mercury that such times can be accurately recorded. As he wrote in his 1716 *Philosophical Transactions* paper:

> I discovered the precise quantity of time the whole body of Mercury had then appeared within the Sun's disc, and that without an error of one single second of time.[11]

However, some care must be taken with the calculations because the movement of Earth in its orbit and its rotation must also be considered over the period of hours involved in the transit. For Venus's spot to cover a solar diameter, it takes almost eight hours. (See Phillips for an introduction to the calculations used when discussing a transit.)

6.5.3 Results

The French astronomer Joseph-Nicolas Delisle (1688–1768) met Halley in 1724 and became a champion for his methods as the latter aged (and died in 1742). In his 1716 paper, Halley discussed suitable sites for observers, and, in May 1760 (so just in time), Delisle published a world map showing regions where the transit could be observed with instructions for observers. As the transit date of June 1761 approached, various scientific organizations began sending expeditions to sites all over the world. (This is a fascinating piece of scientific history, and the entertaining books by Woolf and Lomb relate the full and multifaceted story.) In the end, 120 observations of the 1761 transit of Venus were made at locations spread worldwide by astronomers mostly from France and Britain, but Sweden, Germany, Denmark, Italy, Russia, and Portugal were also represented. (Woolf gives a table with all the details.)

The accuracy of the results varied with the expertise and location of the observers. It became apparent that there are also technical problems. For example, it is difficult to decide exactly when the shadow spot encounters the sun's disc, the so-called black-drop effect. Results at the time were expressed in terms of the solar parallax, and values ranged from 8.28" to 10.60". Since Halley had predicted an error "within the 40th part of one second" this was a disappointing outcome. (An angle is measured in degrees; a degree is divided into 60 minutes and each minute of angle is divided into 60 seconds, or 60 arcseconds.)

For the 1679 transit, Woolf tabulates 150 observers, again spread far and wide across the globe. The results were generally better this time, but examination of the observations and analysis by different astronomers produced inconsistent final values. For example, the eminent French astronomer Lalande claimed the answer was between 8.55" and 8.63". English observations suggested 8.8" and the Swedish Academy settled on 8.5".

Whatever the final outcome details, these calculations based on Halley's ideas finally showed that the earth is around 149 million kilometers from the sun. Setting the scale of the solar system was a major achievement; it is a fine example of how calculations can take simple observations to obtain a physically important parameter. I add **calculation 19, finding the astronomical unit** to my list. It is one of the sad points in physics where we realize that Edmond Halley used calculations to make two great predictions (the return of Halley's Comet and the determination of the solar parallax or Earth-sun distance using a transit of Venus), but did not live to see either of them wonderfully proved correct.

6.6 ROTATING ORBITS

Kepler's first law states that the orbit of a planet is an ellipse with the sun at one focus as shown in figure 6.8 (a). In his *Principia*, Newton showed that his law of gravity, equation (6.1), gives rise to orbits that are conic sections (thus including ellipses). He also looked at the inverse problem: Which forces give an elliptical orbit? He showed that only the inverse square law force, as in equation (6.1), or a force depending on distance linearly, would do the job. (Newton also refined the dynamics of systems comprising two or more bodies by building in center of gravity concepts. See Pask for details.)

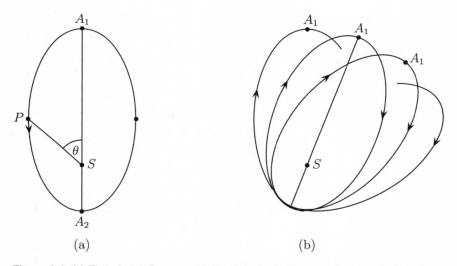

(a)   (b)

Figure 6.8. (a) The planet *P* moves in an ellipse with the sun *S* at one focus. A_1 and A_2 are the orbit apsides. (b) When the force is not an inverse square law force, the orbit does not close but slowly rotates, and there are no fixed apsides. The apsides are the points on the orbit that are nearest and most distant from the sun. *Figure created by Annabelle Boag.*

Newton also wished to find properties of the inverse square law force that would enable him to further support his claim that it represented the force of gravity. This led him to study the behavior of orbits when the force deviated from inverse square form, so instead of r^2 in equation (6.1), we would have r^p. The results are in his *Principia* (Section 9: The Motion of Bodies in Movable Orbits; and the Motion of the Apsides).

Newton considered orbits that are near circular, which is appropriate as the ellipticity of planetary orbits is generally small. In a mathematical tour de force, he showed that unless p is exactly equal to 2, the orbit rotates as shown in figure 6.8 (b). After one rotation of the line from the sun to the planet through 360°, the orbit only returns to its original point when $p = 2$; a line drawn through the apsides rotates, slowly, when p is close to 2. (See Pask chapter 14 for an introduction to this work.) This property of forces and closed orbits is now known as Bertrand's theorem.

θ in figure 6.8 (a) increases by π radians (or 180°) as the planet moves between the apsides, A_1 to A_2, and then moves through another angle π to get back to its starting position. The line of apsides A_1A_2 remains fixed in space, and this is a property of the orbits generated by an inverse square law force. If the angular change in θ varies from π, the apsides move and the line joining them rotates. See figure 6.8 (b). Newton calculated that

$$\text{if } F = \frac{GmM}{r^{3-n}}, \text{ the position angle } \theta \text{ increases by}$$

$$\frac{180^o}{\sqrt{n}} \text{ as the planet moves between apsides.}$$

Obviously, for $n = 1$, the force is inverse square and the angle is 180°. Newton gave examples for other values of n. He then points out that the formula can be used in the inverse way; if the orbit rotates, that will determine n, and then we know that the force producing the orbit varies inversely as r^{3-n}.

6.6.1 Solar System Data

The data for the planets in the solar system reveal that their orbits are closed to a high degree of accuracy, thus giving validity to the inverse square law of gravity. Newton pointed out that the perturbations caused by the other planets will explain any small deviations. In particular, as we saw in section 6.2, Jupiter and Saturn are large planets and will have small but significant effects on the motion of the other planets.

This is a typical example of the work in celestial mechanics over the centuries after Newton. Properties of various planetary and lunar effects were calculated using equation (6.1) as the law of gravity and then any deviations between theory and observations were accounted for. Occasionally there was a crisis, and someone (even the great Euler on one occasion) suggested the deviations could only be reconciled with the theory if the inverse square law failed to be exactly correct. Recall that in section 6.4 we saw that changes to the law of gravity were suggested as a way of accounting for the strangeness of Uranus's orbit before the perturbing planet Neptune was discovered. It was invariably found that difficulties in the complex calculations were the source of the errors, and there was no evidence for interactions between the sun and the planets or between the planets themselves to be any other than those described by equation (6.1).

6.6.2 A Stubborn Discrepancy

Newton's theory of gravity is one of the greatest scientific achievements, linking and explaining phenomena observed on the earth and in the heavens. Naturally, all observations were used to test the theory, and the triumphs were cause for celebration—as with the successfully predicted return of Halley's Comet. However, in the nineteenth century, one small discrepancy between theory and observation refused to go away.

Mercury is the smallest planet. It is closest to the sun, and its orbit deviates most from a circle. As data on the orbit of Mercury increased, it was found that an apsidal line through its orbit rotated through 565" over a century. An evaluation of the effects of the other planets showed the following contributions:

Venus – 280.6"
Earth – 83.6"
Mars – 2.6"
Jupiter – 152.6"
Saturn – 7.2"
Uranus – 0.1".[12]

Each of these figures is reached after a major calculation using perturbation theory (see Parks section 7.6 for an example in modern notation). The major contributions come from the planet nearest to Mercury (Venus) and the most massive planet (Jupiter). Those individual contributions total to 527", leaving the small but significant amount of 38" unaccounted for. It may seem that a rotation of thirty-eight seconds of angle taken over a century is not something to quibble about, but the excellence of the theory-observation verification suggested otherwise.

After his triumph with Neptune, Leverrier was naturally interested in this odd result. In 1849, he pointed out the above anomaly and began work checking the calculations. He tried variations in the masses of the planets, but no change consistent with other results and observations could account for the missing 38". It will come as no surprise that Leverrier then suggested that there was once again an undiscovered planet, this time perturbing the orbit of Mercury. In order to leave the good fit for Venus's orbit unchanged, he suggested that the orbit of the new planet must lie between the orbit of Mercury and the sun. Eventually this undiscovered planet was given the name Vulcan.

This time there was enthusiasm in many places for searches to reveal Vulcan, and several false starts were recorded. (This is a fascinating period in astronomical history, and the articles by Fernie, Morando, and especially Hanson are recommended.) There was no consistent solution to the Mercury rotation problem, and the suggested planet Vulcan remained undiscovered (as it does today!). American astronomers such as Simon Newcomb were involved in the story and checking the calculations. By the twentieth century the famous missing rotation was believed to be 43".

6.6.3 The Mystery Solved

The unexplained rotation of Mercury's orbit is an astronomical fact that does need a change in the law of gravity to be used in its calculation. In one approach, the American astronomer Asaph Hall (1829–1907) used Newton's theory to suggest that the 2 giving the power of the distance r in equation (6.1) should be changed to 2.000000157. That would give the extra 43" of orbit rotation, and the change would only be significant for Mercury as it is the planet nearest to the sun. As it turned out, the law of gravity really does need to be modified, and the effect only shows up strongly for a planet close to the massive sun. However, it is not the modification envisioned by Hall.

Albert Einstein examined the most basic assumptions and ideas on which physics is based, and he was led to his theory of relativity. In 1916, he published *The Foundation of the General Theory of Relativity*, which gives a different approach to the description of gravity. But still, Newton's theory emerges as the first approximation to Einstein's theory. For calculating planetary orbits that first approximation must be modified, and in the approach using forces, equation (6.1) becomes

$$ F \; = \; \frac{GmM}{r^2} + \left(\frac{3GmMh^2}{c^2} \right) \frac{1}{r^4}. \qquad (6.4) $$

There is an additional force term (the second term in equation (6.4)), which depends on the angular momentum constant h, the speed of light c, and the distance to the fourth power r^4. Because c is so large, the magnitude of the correction term is very small.

We no longer have a purely inverse square law of gravity, and as a result, the orbits calculated with it are not closed. Here, then, is the explanation for the anomalous rotation of the orbit of Mercury. (You can even use Newton's theory to calculate the rotation effect—see Pask chapter 14.) Using the relativistically correct equation (6.4) instead of equation (6.1) gives just that missing 43" amount of rotation. According to Einstein's biographer Abraham Pais:

This discovery was, I believe, by far the strongest emotional experience in Einstein's scientific life, perhaps in all his life. Nature had spoken to him. He had to be right. "For a few days I was beside myself with joyous excitement" (Einstein in a letter to Ehrenfest). Later he told Fokker that his discovery had given him palpitations of the heart.[13]

These are the calculations of Newton, Einstein, Leverrier, and a host of diligent theoretical astronomers. The physics and mathematics are brilliant, and the conclusions they lead to are profound. **Calculation 20, rotating orbits** must have a place in the list of great calculations.

6.7 OTHER CONTENDERS

This part of science is full of wonderful stories and marvelous calculations, so keeping to a limited selection has not been easy. In particular, I regret not being able to include two brilliant pioneering efforts. First there is Carl Gauss's 1809 *Theoria Motus*, which describes the mathematical procedures to use for determining the full details of an orbit when just a few observations are known. Gauss used very limited data to find the orbit of the asteroid Ceres and predict where to find it after it was lost behind the sun. Ceres was the "missing planet" between Mars and Jupiter referred to in section 6.4.1. (See the article by Marsden for more information.) Second is the work on the extremely difficult problem of understanding the moon's orbit and drawing up tables of lunar positions. This has continued to be an important enterprise ever since Newton struggled with it, and wonderful advances were made by George W. Hill (1838–1914), one of America's great astronomers. (See Morando for further details.) It has been hard to leave Gauss and Hill off my final list.

Chapter 7

THE UNIVERSE

*in which we see how calculations help
to answer some of the biggest questions of all.*

The final step from chapters 4, 5, and 6 takes us to the ultimate question: What is the nature of the whole universe? Our earth is but a speck in the vastness of the universe, something which became ever more apparent after Galileo saw a new brightness of stars through his telescope. Cosmology is for many people the most inspiring, challenging, and mystifying topic in all of science, and every advance seems to leave us with new questions. Much of cosmology involves extremely technical matters, but in this chapter, I describe five significant calculations that illustrate how the subject progresses without, I hope, being too overpowering for the general reader. Other relevant calculations will be found in the chapters on light (see section 9.6) and nuclear physics (see section 11.7).

The ancient Greeks placed the stars on a crystal sphere defining the outer limits of the universe. But already by the time of Titus Lucretius Carus (around 99 BCE to 55 BCE) there were strong arguments against such a picture. Lucretius wrote a wonderful poem summarizing all knowledge: *De Rerum Natura*, translated as *The Poem on Nature* or *The Way Things Are*. In the poem he writes:

> The universe itself is without limit of any kind,
> For if it had it would have to have an outside.
> Nothing can have an outside unless there is something beyond it
> So the point can be seen at which it ceases to be
> And beyond which the senses could follow it.
> There can be no such point for the whole of creation;
> If one thinks of the whole there can be nothing outside it,

It can have no limit or measure, you could not conceive it.
It does not matter what position you occupy,
Space must stretch an infinite distance in every direction.[1]

Here we have the great dilemma set out. Lucretius uses no mathematics, but he does give a lovely argument involving a spear thrower at the edge of the universe: Does the spear bounce back, or go on forever? Asking about the size of the universe, and about its age, origin, and composition, are among mankind's most profound questions.

7.1 THE DARK SKY

Go outside at night and you may see a sky covered in stars. You might say that the brilliant stars shine out of a great blackness. For some people, including Kepler and Halley, that turned into a puzzle: the question is not about the stars we see, but rather, why is the sky dark around them? The nature of this puzzle and the reactions it provoked are an important part of the history of cosmology. This is a good example of a problem arising from a simple calculation, and it is a problem that is solved by carrying out a series of other calculations. (To explore the full story in detail the books and papers of E. R. Harrison are recommended.)

7.1.1 Posing the Problem

Thirty years after the death of Copernicus, the astronomer Thomas Digges (1543–1595) published *A Perfit Description of the Caelestiall Orbes* in which he further reduced the special nature of mankind's status by placing the solar system amongst an infinite number of stars spreading out into an unbounded space. See figure 7.1. Digges was a modern day Lucretius.

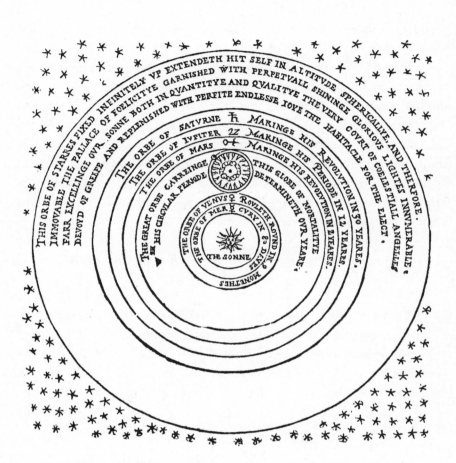

Figure 7.1. The solar system in an infinite universe as proposed by
Thomas Digges in 1576. *From* Wikipedia, *user Paddy.*

Digges obviously realized that there might be problems in assuming
an endless, infinite number of stars, but he concluded that "the greatest part
rest by reason of their wonderful distance invisible unto us."[2]

Kepler reacted to this idea of an infinite universe with terror and
believed that it implied the whole "celestial vault would be as luminous as
the Sun" and so "this world of ours does not belong to an undifferentiated
swarm of countless others."[3] Thus the puzzle and a possible solution began
to take shape: the universe cannot be infinite because that implies a bright

night sky (Kepler), unless in some way the most distant stars are "invisible unto us" (Digges).

Edmond Halley recognized the problem and gave it a more mathematical basis, which we can set out formally as follows. Suppose we assume the space around us is divided into spherical shells each filled uniformly with stars, n per unit volume, as in figure 7.2. Let a star contribute brightness B (in some measure) which will reduce inversely with the square of the distance from the observer according to the well-known laws of optics. For the shell of thickness dr at distance r the volume is $\pi r^2 dr$ and so the brightness at the observer B_{obs} from a total sphere of radius R is given by summing or integrating over all the shells:

$$B_{obs} = \int_0^R (n\pi r^2)\left(\frac{B}{r^2}\right) dr = n\pi B \int_0^R dr = n\pi BR. \quad (7.1)$$

Thus the r^2 in the volume cancels with the $1/r^2$ in the optical factor, and all shells contribute in the same way; the total brightness over all the spherical shells is proportional to the radius R of the whole universe. Now, if R is infinite, the brightness of the sky would become infinite, and, as Kepler feared, we would be burnt up in the hot radiation falling on the earth.

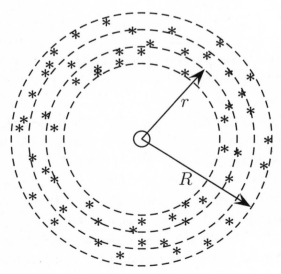

Figure 7.2. Spherical shells of stars shining onto a central observer. A typical shell is at radial distance r, and the final shell has radius R that tends to infinity in an infinite universe. *Figure created by Annabelle Boag.*

The German astronomer Heinrich Olbers (1758–1840) wrote later about this dark-sky puzzle, which has become known as Olbers's paradox—named rather unfairly to those who studied the puzzle before Olbers's time.

It is interesting to note that one simple calculation and the simple observation that the night sky is mostly dark force us to confront some of the great questions about our universe.

7.1.2 The Solution

Halley followed Digges, suggesting that light from the most remote stars would not be detectable by us for some unspecified reason. In 1744, the Swiss astronomer Jean-Phillippe Loys de Chéseaux gave careful calculations about star numbers and effects but still concluded that the earth would receive light far brighter than sunlight. Chéseaux (and Olbers) suggested that the solution was absorption of light by interstellar matter. Herschel later pointed out that the interstellar material would become heated and hence radiate, so the overall problem remained.

One of the first things to realize is that the radius R in equation (7.1) should really be the "look-out limit." As we move out from the observer, the number of stars increases and gradually fills the sky so that eventually the light from stars further out is blocked by those nearer to the observer. (The analogy of looking through a forest is often used; after a certain distance all further tree trunks will be blocked from sight.) This limit can be calculated and is still very large—about 10^{23} or 10^{24} light years according to Edward Harrison.

There are two other numbers that must be noted. First, the speed of light is finite (300,000 km/sec), so we must consider light that has had time to reach us. Second, stars have a finite lifetime, say 10^{10} years. Harrison claims that Lord Kelvin was the first person to bring all of these sorts of facts together in calculations reported in a paper published in 1901. Harrison goes on to describe ideas set out by Edgar Allan Poe that also make sense of the dark-sky mystery. The final viewpoint on the matter, according to Harrison, is summarized as follows:

Darkness of the night sky is due not to the absorption of starlight, not to hierarchical clustering of stars, not to the finiteness of the universe, not to expansion of the universe, and not to many other proposed causes. The explanation is quite simple and can be stated in various equivalent ways. Because of the finite luminous age of stars and the finite speed of light, the number of visible stars is too few to cover the entire sky; most stars needed to cover the sky are so far away that their light has not reached us; the light travel-time from the most distant stars is greater than their luminous lifetime; . . . Why is the sky dark at night? Because starlight is too feeble to fill the dark universe.[4]

The dark-sky problem is intriguing, and despite Harrison's views (which are often taken as definitive), it still stirs up controversy. The dark-sky problem played an important role in making cosmologists consider the universe and its properties and is a worthy addition to my list is **calculation 21, why the night sky is dark**.

7.2 WHAT SORT OF UNIVERSE

When we look up into the sky, we only see a miniscule part of the universe, and we are drawn to certain particular visible structures such as the planets and the Milky Way. To make progress on a larger scale, we need to find a guiding principle. The one generally assumed is the cosmological principle: the universe is homogeneous and isotropic—it is the same everywhere and in any direction we care to look. Obviously at first sight the universe does not appear to be the same in every direction, but that only emphasizes the enormous scale on which we must operate. We know there is a mix of stars, galaxies, and all sorts of strange objects, and we assume that mix continues equally everywhere in the universe when we extend our considerations to the largest scales. Observational evidence and astronomical surveys support the cosmological principle.

Using the cosmological principle allows us to use observations to infer the most general properties of the universe. How that is done using the combination of observation and theory—the use of calculations to make sense of data—is the subject of this section.

This section provides a nice example of three important uses of calculations:

1. To help deal with experimental data and their presentation
2. To deduce new results from a given theory
3. To calculate values for physical parameters which may be used to describe the universe and promote further experimental work

7.2.1 Calculating with Observational Data: Hubble's Discovery

Studies by early workers such as V. M. Slipher and C. Wirtz indicated that spiral nebulae were moving away from us at great speeds, and those speeds seemed to be correlated with the brightness of the spiral nebulae. The speeds are calculated from the amount of redshift in the observed spectral lines. Concepts and observations were developed that allowed properties such as brightness and its variations to be used to calculate estimates of the distances to those nebulae. (For a simple introduction to these matters see the books by Coles and Weinberg.) The stage was set for one of the most famous papers in cosmology: *A Relation between the Distance and Radial Velocity among Extra-Galactic Nebulae*, published by Edwin Hubble in 1929. Hubble presented his data in a diagram as in figure 7.3.

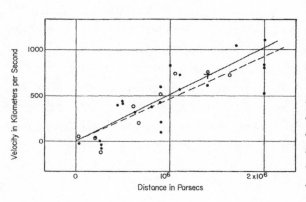

Figure 7.3. Hubble's 1929 plot of velocity versus distance for extragalactic nebulae. *From Edwin P. Hubble, "A Relationship between Distance and Radial Velocity among Extra-Galactic Nebulae,"* Proceedings of the National Academy of Science *15, no. 3 (March 15, 1929).*

Hubble's results say that the observed nebulae are receding from us with velocities that increase with the distance. This result has now been generalized and interpreted using the cosmological principle to give a result central to all cosmology:

> the universe is expanding; if observed from any point, elements in the universe are receding from that point with speeds which are increasing with the separation distance.

But Hubble went further than that with his statement: "the results establish a roughly linear relation between velocities and distances among nebulae."[5] This may well be the most significant result in cosmology, and it is now written as

Hubble's law: $V = Hr$

or $V = H_0 r$ if H is time dependent and now $H = H_0$. (7.2)

Here V is the speed of recession of the object being observed at distance r. Naturally, such a basic finding about the absolute nature of the whole universe has led to an enormous effort to verify and extend Hubble's original results. (For a detailed, comprehensive study see the 2010 review by Freedman and Madore). A more recent diagram of the data is shown in figure 7.4.

This is a good example of how a calculation (here a data fit and suggesting a simple linear relationship) identifies an important parameter. Hubble's law as stated in equation (7.2) identifies H_0 as a (if not the) key parameter in cosmology. I return to this point in section 7.2.3.

(A historical aside: there is some debate about who deserves credit for the work discussed in this section; as is often the case, several people, probably independently, came to conclusions about the expanding universe similar to those reached by Hubble. See the letter by Way and Nussbaumer for a summary and references, and the later response by Livio.)

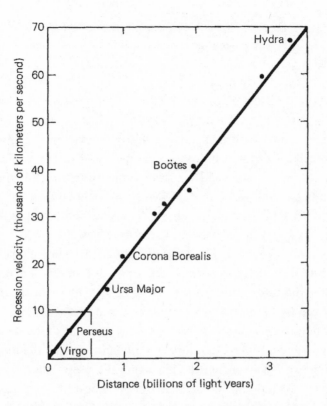

Figure 7.4. More recent data supporting Hubble's law. The small box at the origin indicates the extent of Hubble's original data. *Reprinted with permission, © Cambridge University Press, from E. R. Harrison,* Cosmology: The Science of the Universe, *2nd ed. (Cambridge: Cambridge University Press, 2000).*

7.2.2 Calculating from Einstein's General Relativity Equation

In Isaac Newton's theory of dynamics, bodies interact via forces, and their movements may be measured relative to a given coordinate system or reference frame. In particular, the force controlling the motion of large bodies in space is gravity acting over large distances according to equation (6.1). Newton was concerned about the stability of the universe, and in his *Principia*, he resolved the stability difficulty with a view that

lest the system of fixed stars, by their gravity, fall on each other mutu-
ally, he [God] hath placed those systems at immense distance from one
another.[6]

Interestingly, the idea of the collapse of a finite universe was considered by Lucretius in his *De Rerum Natura*.

It was not until 1916 that suitable ways to tackle these problems were
set out by Albert Einstein in his general theory of relativity. The Newtonian concepts were replaced by a strange theory in which bodies change
the properties of the space through which they move thus affecting how
they move. In John Archibald Wheeler's famous words "space-time tells
matter how to move; matter tells space-time how to curve."[7] Einstein's
equations can be used to explore the properties of the universe on the very
largest scale in which the cosmological principle holds. Instead of considering the discrete entities, like stars, which we observe, the universe is
characterized by the properties of density ρ and pressure p corresponding
to a continuum. Of course the observed entities lead to the density when
suitably averaged. If these assumptions are built into the theory, the result
is the Robertson-Walker metric expressing the properties of space-time.
This metric contains a scale factor $R(t)$, which describes how the spacing
varies in any local coordinate system in use at time t varies. Bodies may
be referred to—such as a Cartesian grid of coordinates, as usual—but over
time the size of that grid changes. (This is described in every book on cosmology; I recommend the treatment in the books by Lambourne—see the
very clear chapter 8—and Roos.)

Using the continuum density and pressure model in Einstein's equations leads to the Friedman equations for the scale factor $R(t)$:

$$
\left[\frac{1}{R} \frac{dR}{dt} \right]^2 = \frac{8\pi G}{3} \rho - \frac{kc^2}{R^2},
$$

$$
\frac{1}{R} \frac{d^2 R}{dt^2} = -\frac{4\pi G}{3} \left(\rho + \frac{3p}{c^2} \right). \tag{7.3}
$$

In these equations, G is the usual Newtonian gravitational constant, and k is a factor related to the spatial geometry involved. (At present it seems likely that k may be zero.) Readers not familiar with the mathematics involved should simply note that the first equation tells us the scale factor in terms of the density ρ (if k is indeed zero) and that the scale is increasing. The second equation tells us about the rate of change of the scaling factor: Is it a steady change (increase or decrease) or is the rate of change itself varying?

The calculation in this case has taken us from Einstein's equations with certain input assumptions to new equations for the scale of the universe.

7.2.3 Results

We can now link the two calculations (one on Hubble-type experimental data, and the other on Einstein's equations under certain assumptions) because the Hubble factor $H(t)$ and the scale factor $R(t)$ are linked by

$$ H(t) \; = \; \left[\; \frac{1}{R} \frac{dR}{dt} \; \right] . $$

Using the above calculations and available data, we can come to the remarkable conclusion that

> the Universe has a nearly flat spatial geometry with $k = 0$ and a total density that is close to 1×10^{-26}kg m^{-3}. Such a universe originated with a big bang . . . and has an expansion age of about 13.7 billion years.[8]

Such a staggering result is why I add **calculation 22, state of the universe** to my list.

7.2.4 A Weird Twist to the Story

Recall that Lucretius and Newton were both concerned by the way gravitational effects would cause matter to move closer together and produce a collapsing universe. The second Friedman equation (7.3) tells us that the second derivative of the scale factor is negative, implying that the rate of expansion of the universe should be slowing down, in keeping with the contracting effect of gravitational forces. However, observations (for which the astronomers Saul Perlmutter, Adam Reiss, and Brian Schmidt were awarded a Nobel Prize in 2011) indicate that the rate of expansion is actually increasing.

The suggested solution to this dilemma is to postulate the existence of something called dark energy, which contributes a density ρ_{de} and has the weird fluid property that its pressure is negative so that the right-hand side of equation (7.3) becomes positive. Dark energy is somehow driving an acceleration in the rate at which the universe is expanding. It acts something like a repulsive gravitational force. Current data suggest that at this time, dark energy contributes well over 68 percent to the density of matter in the universe. This is a weird and disturbing result. The "dark" refers to the fact that this dark energy has not been seen (detected) in other ways, and here is an incredible mystery for future cosmologists to come to grips with.

I must point out that this is connected with one of those wonderful stories in science: Einstein's worries over his cosmological constant Λ. In his early work Einstein introduced a constant Λ into his theory to satisfy a requirement that the universe could be static, as was believed to be the case at the time. He did not like the introduction of what seemed to be an arbitrary constant, and at various times Λ has disappeared. It now seems to have resurfaced in the guise of dark energy. (Interested readers might consult "Lambda: The Constant that Refuses to Die," an entertaining review by John Earman.)

7.3 CHEMICAL ELEMENTS, THE UNIVERSE, AND US

I included this section for four reasons: it relates to why we can actually exist, it shows us some very clever and innovative thinking, it involves a remarkable prediction made by a remarkable astrophysicist, and it introduces the fascinating subject of a link between humanity and the nature of the universe.

In the previous section, we saw how the nature of the universe has been revealed using a mixture of theory and observation. It is now widely accepted that the universe began with the big bang. In the earliest times of the universe there was a hot, dense mixture of photons, neutrons, protons, and electrons. The theory explains how the universe cooled and expanded, and how gravitational forces eventually pulled matter into clumps that became stars and galaxies. But there is no mention of the essential chemical elements that go to form us and our world. Where did they come from? The answer is one of the most extraordinary stories in all of science. It is also a story that inspires philosophical thoughts about the universe and our place in it.

7.3.1 Review

I will deal with topics in atomic and nuclear physics in later chapters, but a few points are needed here. An atom consists of a central nucleus about which orbits negatively charged electrons. The nucleus comprises positively charged protons (supplying the attractive force to keep the negatively charged electrons around it) and neutrons. The strong nuclear force holds the protons and neutrons together to form the nucleus. It is the production of nuclei which concerns us here since electrons are attracted later to form atoms.

The simplest element is hydrogen, denoted by H^1, which has one proton as its nucleus and one electron orbiting that proton. We can also add neutrons to produce heavy hydrogen: the deuteron D or H^2 has one proton and one neutron in its nucleus; adding one more neutron gives the triton or H^3.

The elements build up, with more and more protons in the nucleus and with corresponding numbers of electrons in orbits around the nucleus. Thus we get the nuclei for

helium: He^3, two protons, one neutron; He^4, two protons, two
   neutrons;
lithium: Li^5, three protons, two neutron; Li^6, three protons, three
   neutrons;
beryllium: Be^7, four protons, three neutrons;

and so on through the familiar periodic table of elements. Increasing the number of protons by one each time, we find the next in line to be boron, carbon, nitrogen, and oxygen.

Notice that the number of neutrons may be varied, but in each case the number will affect the stability of the nucleus. (There is more on that topic in chapter 11.) Of great importance is the fact that the Li^5 and beryllium nuclei are not stable, and after forming, they quickly split apart again.

The light elements, hydrogen and helium, are most abundant in the universe with all the other elements occurring thousands or more times less often. Still, those more complex elements are needed to form us! Where they come from is a vital question for science to answer.

7.3.2 The Fusion Process

In the fusion process, two nuclei join together to form a new type of nucleus. (More details of this process will be given in sections 11.6.1 and 11.7.) This process occurs with the lighter nuclei up to iron, Fe^{55}, which contains twenty-six protons. Energy is released during the fusion process, and I will return to the importance of this in chapter 11. (For elements beyond iron, it is the fission process that dominates nuclear physics.) Here is an example of a chain of fusion processes taken from the notebook of Fred Hoyle[9] (more of whom in a moment):

$$O^{16} + He^4 \rightarrow F^{19} + H^1$$
$$F^{19} + H^1 \rightarrow Ne^{20} + h\nu$$
$$Ne^{20} + He^4 \rightarrow Na^{23} + H^1$$
$$Na^{23} + H^1 \rightarrow Mg^{24} + h\nu$$

Energy released can be in the form of kinetic energy or photons (shown as $h\nu$ in the above).

It would appear that fusion can take us from the original few big bang constituents to the whole range of elements. Certainly helium was formed and is abundant in the universe. But there is an interaction problem when we consider the heavier elements. The nuclei are positively charged and repel each other with the electrical Coulomb force. One nucleus must get close to another in order for the very short-range, strong nuclear force to take over. In other words, nuclei must overcome the "Coulomb barrier" if fusion is to take place. This was extremely unlikely to happen in the conditions prevailing soon after the big bang.

The solution to the interaction problem is found by looking at the interior of stars that are formed and collapse according to the action of the long-range gravitational force. Eventually the stellar interiors become incredibly hot and dense, with enormous pressures created to counter gravitational collapse. It is in these circumstances that nuclei can be propelled over or through the Coulomb barrier and fusion occurs. We have a process called stellar nucleosynthesis in which the heavier nuclei are built up from the lighter ones beginning with hydrogen and helium.

A pioneering paper on the subject was published in 1954 by Fred Hoyle, and the great *Synthesis of the Elements in Stars* was published in 1957 by Margaret Burbidge, Geoffrey Burbidge, William Fowler, and Hoyle. These papers examined chains of reactions (like those set out in Hoyle's notebook) producing the elements and calculated how they could operate in conditions likely to be found in stellar interiors. These are truly landmark papers and calculations.

A final step is needed: How do the elements created inside stars get to form us? The answer is that stars can explode, and material from them is scattered throughout the universe. That matter is then drawn into clumps that can form planets like Earth. Then there can be objects on Earth like us. "We are all made from stardust" is a wonderful way to say it. Surely that is one of science's most fantastic and beautiful stories!

7.3.3 A Momentous Hiccup

The whole building up process, starting with hydrogen H^1 and helium He^4, sounds brilliant, but there is an almost immediate problem: the beryllium nucleus Be^8 needed to move up to carbon C^{12} is unstable. Unless something special occurs, no carbon can be produced. But carbon is produced (as our bodies testify!), and Hoyle found the mechanism. There can be a resonance situation in which, despite the rapid decay of beryllium, the mechanism still allows carbon to form. In his 1954 paper, Hoyle writes the process (with the alpha particle notation α for He^4)

$$\alpha + \alpha \leftrightarrow Be^8 \qquad Be^8 + \alpha \rightarrow C^{12} + \gamma$$

The second reaction produces the gamma ray γ. But hidden in there is the vital step: the carbon nucleus must be able to assume a most particular state.

An aside is required: The nucleus is a mixture of protons and neutrons held together by the strong nuclear force, which overcomes the protons' electrical repulsion forces. According to the appropriate mechanics (that is quantum mechanics, to be introduced more fully in chapter 10), the protons and neutrons form a lowest energy configuration called the ground state. However, if energy is added, they may take up new configurations called excited states.

Now, back to the fusion process that gives us carbon. Hoyle calculated that to form the resonance situation for the production of carbon, an excited state of the carbon nucleus must be involved. This excited state must have an energy above the ground state of about 7.65MeV. (MeV is a unit of energy in nuclear physics. Appendix 3 of Hoyle's 1975 textbook provides a very clear explanation of this work.) Since carbon does exist, Hoyle believed that this particular excited state of the carbon nucleus must also exist. He took his prediction to William Fowler and colleagues who were persuaded to try the necessary experiment. The excited state did indeed exist, just as Hoyle had predicted. This prediction must rank as one of the most remarkable and profound in all of physics.

Solving the problem of carbon production was a brilliant achievement by a remarkable man. Fred Hoyle (1939–2001) was born in Yorkshire, England, and studied mathematics and physics at Cambridge University. He went on to become an outstanding astrophysicist, contributing to many areas of the subject. (For some unknown and shameful reason, he was never awarded the Nobel Prize, although his collaborators, like William Fowler, were.) Hoyle was a proponent of the steady state theory of the universe. He disliked the expanding universe model we accept today, and he scornfully called it the big bang model! Hoyle's sarcasm has been lost, and the mocking name he invented remains in use today. His autobiography, *Home is Where the Wind Blows*, is a captivating read; his version of the carbon story is given in chapter 18, "An Unknown Level in Carbon 12."

It is impossible not to put **calculation 23, Hoyle makes carbon** in my list of important calculations.

7.3.4 Mankind and the Universe

Recall that Copernicus moved the earth from the center of the universe and that gradually we have found that there is nothing special about our Earth or its position in the universe. We have been led to the cosmological principle as discussed in section 7.2.1. Nevertheless, it seems to be part of human nature to return to the so-called big questions like: How and why did it all begin? What is the point of life? What is the meaning of life? What is our place in the whole scheme of things, in the whole universe? While many scientists would say these are not questions for science to answer, in recent times, others have taken aspects of them more seriously. The result has been to add the anthropic principle to the cosmological principle introduced in section 7.2. (Anthropic from the Greek *anthropos*, human being.)

Brandon Carter introduced the anthropic principle (AP) in 1974. In its weak form (WAP) it states that

> what we can expect to observe must be restricted by the conditions necessary for our presence as observers.

This seems noncontroversial; since we exist, the conditions in the universe must allow for that. The existence of the necessary excited state in the carbon nucleus as predicted by Hoyle is often cited as an example of the WAP in action.

The strong anthropic principle (SAP) states that

> the universe necessarily has the properties requisite for life—life that exists at some time in its history.

That word *necessarily* makes this a much more controversial statement. Essentially we can read SAP as saying that the universe must have observers and so must have conditions suitable for their existence. In recent years, there has been much discussion about the "fine-tuning" of the physical parameters, like force strengths, which seem to be necessary for life (as we know it) to exist. It is sometimes claimed that the required coincidences are exquisitely precise and tuned to one another. Of course, it is not too big a step to invoke the idea of a god who carefully designed the universe, and the whole debate extends into a number of philosophical and religious issues.

(The literature on this subject is large; the interested reader might start with Paul Davies's book *The Goldilocks Enigma*—Goldilocks, remember, wanted everything to be just right. For a brief introduction see chapter 8 in Harrison's *Cosmology*. An insightful and entertaining opinion of anthropic principles is given by mathematician-philosopher-writer Martin Gardner. He calls the final anthropic principle (FAP) the "completely ridiculous anthropic principle"—you can work out the abbreviation for yourself and get an idea of Gardner's take on the subject!)

To conclude, it is appropriate to return to Hoyle. His work on carbon predates the anthropic principle, and certainly in his 1954 paper there is no mention of the necessity of carbon for life. (The paper "An Anthropic Myth" by Helge Kragh provides an extended discussion.) As well as using the excited state of carbon, Hoyle noted the precise energy level of a state in oxygen which this time acted in the opposite sense; if the state in oxygen was a little different, all the carbon would quickly convert to oxygen. He wrote:

> The positioning of these levels [in carbon and oxygen] depends on the electrical repulsion between protons, and the strength of the nuclear forces that bind protons and neutrons within the nuclei. Change those two opposing effects only slightly, and the levels in C^{12} and O^{16} could be changed by an amount that would produce a world essentially without carbon, and hence without life as we know it.[10]

Certainly Hoyle found a reason for some very careful tuning of physical parameters. In his autobiography he asks:

Was the existence of life a result of a set of freakish coincidences in nuclear physics? . . . Or is the universe teleological, with the laws deliberately designed to permit the existence of life, the common religious position?[11]

Debates related to **calculation 23, Hoyle makes carbon** are set to continue for a very long time.

7.4 WHAT IS THE MATTER?

In section 7.2.2, we saw that the average density of matter in the universe is the key parameter in determining its nature—whether it expands or contracts, for example. Hence discovering how much matter is distributed throughout the universe is a central problem of cosmology. We already saw that the strange dark energy contributes about three-quarters of the density to be used in equation (7.3), leaving only about 25 percent of "ordinary matter." However, the peculiarities in the matter question do not stop there. As observational techniques were refined in the twentieth century, a new puzzle arose.

7.4.1 Large-Scale Motion in Galaxies

Gradually it became possible to observe galaxies, their motion, and the motion of their component stars and gases. These are large-scale gravitational phenomena, and so Newtonian theory can be applied. Some of the earliest studies by Jan Oort in 1932 and Fritz Zwicky in 1933 already showed that matching observations and dynamical theory was not straightforward. In his 1937 paper, Zwicky found that different methods failed to give consistent values for the masses of galaxies (nebulae).

The work of astronomer Vera Rubin provides a good example of how the field has progressed. Vera Rubin's life and achievements illustrate the struggles women faced in order to succeed in the field of astronomy and how someone can be an important scientist as well as a wife and proud mother. (Her inspirational story is told in the biographical essay by Robert Irion.) In a (now) amusing sign of the times, the Washington Post headlined a story in 1950 about Rubin with "Young Mother Figures Center of Creation by Star Motion." Remembering this, in 1993 when President Clinton recognized her brilliant career, some of Rubin's friends suggested the headline: "Old Grandmother Gets National Medal of Science"! The famous George Gamow supervised Rubin's PhD studies. When she visited him at the Applied Physics Laboratory at John Hopkins University, their meeting had to take place in the lobby because women were not allowed in the laboratories. Rubin was a pioneer; in 1965, she was the first woman allowed to observe at the Palomar Observatory in California.

Starting in 1965, in collaboration with Kent Ford, Vera Rubin made extensive measurements of motion in the Andromeda, M31 galaxy. Later work extended to other galaxies. (I recommend Rubin's articles in *Science*, *Scientific American*, and *Physics Today* for an introduction to this work and the story of its progress and impact.) The result was observational data on the rotational velocity of stars at different distances from the galaxy center. Many results for different galaxies were accumulated, and figure 7.5 shows two examples taken from Rubin's 1983 *Science* paper.

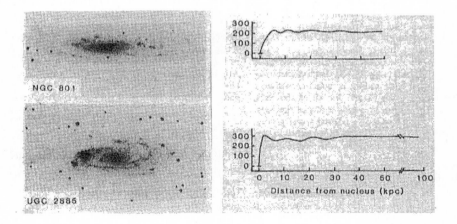

Figure 7.5. Rotation curves for components of the galaxies NGC 801 and UGC 2885 as a function of the component distance from the galaxy center. *Reprinted with permission of AAAS, from Vera C. Rubin, "The Rotation of Spiral Galaxies,"* Science *220, no. 4604 (1983).*

Notice that the velocity of the stars remains the same as we move away from the galaxy center. It is almost as if we had discovered that Jupiter, Saturn, and Uranus all travel with the same velocity. There is something extremely troubling about that result, as I explain below with the relevant calculations, and it led Rubin to give this remarkable summary for her paper:

> There is accumulating evidence that as much as 90 percent of the mass of the universe is non-luminous and is clumped, halo-like, around individual galaxies. The gravitational force of this dark matter is presumed to be responsible for the high rotational velocities of stars and gas in the discs of spiral galaxies.[12]

(Of course, Rubin gave this 90 percent figure not knowing that a very large contribution to the universe was yet to be discovered in the form of dark energy as mentioned above in section 7.2.4.)

There were also results for the motion of several galaxies, or clusters of galaxies, which gave strange results when the virial theorem was used to analyze them. To make clear the magnitude of the effects under discussion, consider this breathtaking start to the 1974 paper by Ostriker, Peebles, and Yahil:

> There are reasons, increasing in number and quality, to believe that the masses of ordinary galaxies may have been underestimated by a factor of 10 or more. Since the mean density of the universe is computed by multiplying the observed number density of galaxies by the typical mass per galaxy, the mean density of the universe would have been underestimated by the same factor.[13]

The crucial parameter—the density of the universe—used so far might be wrong by a factor of ten!

7.4.2 How It Works

It is time for one of those calculations linking observations and theory.

We can understand the source of the above conclusions by using two simple results from dynamics. First, we need to know the force exerted on a body of mass m in a spherically symmetrical mass distribution. In his *Principia*, Isaac Newton calculated that if the body is at a distance R from the center, then it experiences a gravitational force equivalent to a mass $M(R)$ placed at that center, where $M(R)$ is the sum of all the mass in the spherical distribution out to that radius R. A key point is that the mass outside the sphere of radius R exerts no force on the body—all the different contributions average out to zero. The force is given by equation (6.1) after substituting $r = R$ and $M = M(R)$.

Second, we need to know the dynamical law that force equals mass times acceleration; for the body rotating around the mass distribution center with radius R that acceleration is $V(R)^2/R$ where $V(R)$ is the body's velocity at radius R.

Putting those two facts together gives

$$\frac{GmM(R)}{R^2} = \frac{mV(R)^2}{R},$$

(7.4)

so $\qquad M(R) = \frac{RV(R)^2}{G}$ or $V(R) = \sqrt{\frac{GM(R)}{R}}.$

Thus the observed velocities $V(R)$ are directly related to the mass distribution $M(R)$.

One case is very simple. Suppose the galaxy mass distribution ends at a radius R_G and $M(R_G) = M_G$, then for bodies at a distance $R > R_G$ equation (7.4) gives

$$V(R) = \sqrt{\frac{GM(R_G)}{R}} = \sqrt{\frac{GM_G}{R}},$$

so $\quad V(R)$ is proportional to $\dfrac{1}{\sqrt{R}}.$

This means that for a galaxy of size R_G and mass M_G, the dust and few stars out at the edge of the galaxy should have velocities which decrease like the square root of their distance from the galaxy center. This is also the case for planets orbiting the sun; the outer planets are very slow compared with the inner ones as the $1/\sqrt{R}$ indicates. (In fact this is also what Kepler's third law tells us.) But Vera Rubin did not find that type of velocity decrease; she found that the velocities remain constant.

Generally, equation (7.4) tells us that if $V(R)$ is a constant, as the observations suggest, $M(R)$ must be proportional to R. We cannot simply use the galaxy mass M_G found by adding up the masses of its constituent stars; the mass continues to increase as R increases.

Thus the galaxy masses required to fit the observed component velocity patterns do not correspond to the luminous matter. There must be some other "dark matter." As Rubin said about her observation, "this dark matter is presumed to be responsible for the high rotational velocities."

(Of course, the mathematics must be done for galaxies that are not spherical, like the common spiral galaxies, and readers may consult the chapter by Burbidge and Burbidge and the paper by Einasto for the technical details.)

Despite all the complications, the conclusion is clear: galaxies contain a great deal more mass that the visible components suggest, and they seem to have an extensive halo of dark matter around them. I define **calculation 24, galaxy rotation and dark matter**.

7.4.3 Today

Over fifty years have passed since the pioneering work of Oort, Zwicky, Rubin, and many others, and a vast amount of new work has been done on galaxy rotations and the need for dark matter to exist. But the conclusion remains the same. (For example, see the 2009 *Nature* review by Caldwell and Kamionkowski which is headed: "Observations continue to indicate that the Universe is dominated by invisible components—dark matter and dark energy." A typical modern view is given in the Roos textbook, chapter 9, "Cosmic Structures and Dark Matter.") The results of **calculation 24, galaxy rotation and dark matter** remain unchallenged and is a profound contribution to our knowledge of the universe.

Two questions may immediately come to mind. First, if dark matter (and energy) does exist, what is it? There is no definitive answer to that question in 2014. Second, is there a need to modify the underlying theory? The evidence for dark matter relies on the use of Newtonian mechanics, and there have been suggestions that modifications must be made. Yet, to date, there is no evidence that dark matter appears only as a failing of Newtonian mechanics.

After over forty years of personal effort, Vera Rubin could still write in her 2006 review:

> What's spinning the stars and gas around so fast beyond the optical galaxy? What's keeping them from flying out into space? The current answer is, "Gravity, from matter that has no light."[14]

At least dark matter (27 percent) and ordinary matter (5 percent) have normal gravitational properties. The other 68 percent of the universe in dark energy is even weirder (see section 7.2.4): it gives a "repulsive gravitational effect"!

7.5 ESCAPING GRAVITY AND MAKING BLACK HOLES

The force of gravity dominates our life on Earth and is often the factor deciding what is and what is not possible. For example, we must consider gravity when we ask whether humans can ever escape from the earth. In our lifetimes, we have seen truly amazing answers to that question.

Isaac Newton wrote a short book entitled *The System of the World*, which gave a simpler "popular" introduction to his *Principia*. In this book, he describes a thought experiment involving cannon balls shot off from a cannon on top of a mountain. As the muzzle speed is increased, the cannon balls travel further and further before they fall back to Earth. If the speed is great enough, a cannon ball "even might go quite round the whole Earth before it falls." Thus he explains the possibility of the moon's orbit. However, Newton goes further: the muzzle speed is increased "lastly, so that it might never fall to Earth but go forward into the celestial spaces, and proceed in its spaces in infinitum."

The minimum speed needed to escape from Earth and go into the "celestial spaces" is today called the escape velocity v_{esc}. To find it, consider a body of mass m moving with speed v at distance r from the earth so that it has energy E:

$$E = \text{kinetic energy} + \text{gravitational potential energy}$$
$$= \tfrac{1}{2}mv^2 - \frac{GmM}{r}. \tag{7.5}$$

In these equations, G is the usual gravitational constant, and here M is the mass of the earth. The condition for a velocity v to exist no matter how far away the body is from the earth (r becomes infinitely large, so the potential energy reduces to zero) is that E must be positive. The limiting case, when the speed is just enough for the body to keep going forever, will be when $E = 0$. Then we get the escape velocity at the surface of the earth, where $r = R_e$, according to equation (7.5) as

$$0 = \tfrac{1}{2}mv_{esc}^{\;2} - \frac{GmM}{R_e} \qquad \text{or} \qquad v_{esc} = \sqrt{\frac{2GM}{R_e}}. \tag{7.6}$$

At the surface of the earth, v_{esc} is 11.2 km/sec. An object moving upward with this speed can escape from the gravitational pull of the earth.

7.5.1 Twentieth-Century Exploits

The ideas may have been clear a long time ago, but it was not until the twentieth century that mankind developed the technology to make space travel and exploration a reality. Rather than shoot up vehicles with speeds greater than the escape velocity, engineers built multistage rockets and launched satellites into orbit around the earth. The calculations showed how to optimize the rocket design and how to position or adjust the satellites' orbits to give desired surveying and communication properties. Eventually humans were sent into orbit around the earth, still bound to the planet but now far above its surface. We may have become blasé about rockets and space vehicles, but the achievements have been monumental. Two examples illustrate these achievements and indicate the massive calculations involved.

On July 20, 1969, Americans Neil Armstrong and Buzz Aldrin stepped onto the moon, and man had escaped Earth. A now-famous diagram of the way this was achieved is shown in figure 7.6. There are some vital steps: a rocket must carry the astronauts into orbit around the earth; the space vehicle must then take the best path to the moon as it is influenced by gravitational forces of both the earth and the moon; the vehicle must remain in orbit around the moon and a landing craft must descend to the lunar surface; the process must then be done in reverse order and the astronauts returned safely to Earth. Notice that the journeys between Earth and the moon involve motion described by the three-body problem, one which has caused headaches ever since Newton first battled with it.

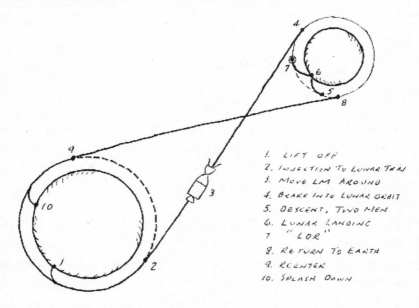

Figure 7.6. The main steps in the process of visiting the moon.
From NASA, John C. Houbolt.

One of the major debates for the mission designers concerned the actual process for landing on the moon. Should the vehicle that goes away from the earth land on the moon, or is some more elaborate scheme required? After considering various factors (such as calculating the fuel needed to take a rocket down to the lunar surface and—more importantly—escaping the moon's gravity again) the lunar-orbit rendezvous concept championed by engineer John C. Houbolt (1919–2014) was chosen. (See the book by Hansen for the story of the debate.) The space vehicle orbits the moon while a module peels off and descends to the lunar surface. On return, the module must rendez-vous with the orbiting parent vehicle. Imagine the calculations that went into designing such a scheme! In fact, Armstrong and Aldrin landed some miles from the targeted landing site, but thankfully the fuel reserves were as planned.

It has now been discovered that spacecraft orbits around the moon cannot be calculated by assuming the usual gravitating sphere model. There are massive dense bodies—"mascons"—embedded in the moon's crust, and they cause significant perturbations to orbits, especially the low-lying ones. In fact, there appear to be just four "frozen orbits" at inclina-tions 27°, 50°, 76°, and 86°; other orbits will be perturbed and wander off course. (See the article by Konopliv.)

The second example of space exploration is the Voyager program. The space probes Voyager 1 and Voyager 2 were launched in 1977 and have become the first man-made objects to set off for interstellar space, thus escaping even the gravitational pull of the sun. The launch dates were chosen so that the probes could take a tour of the outer planets and provide the first close-up reports and extensive data on them and their moons. The orbits are shown in figure 7.7.

The calculations of the spacecraft paths have to take account of gravi-tational pulls by the planets to be visited in order to determine how visiting these planets can be done in an optimum manner. (For details, the textbook by Barger and Olsson is a good place to start, and Roy provides a com-prehensive account of orbital dynamics.) One intriguing result from these calculations is that a spacecraft can gain a "gravity assist" if it flies by a planet in a suitable orbit so that it gains speed as it leaves the region of that planet (see Barger and Olsson, and the tutorial by Van Allen).

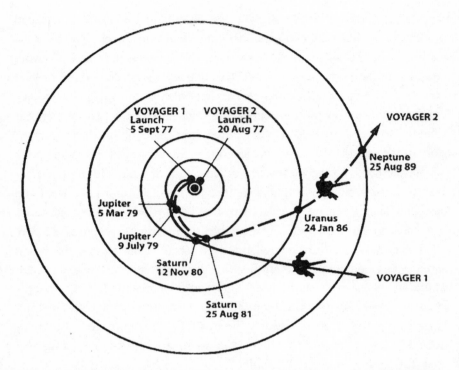

Figure 7.7. Paths in space for the Voyager 1 and Voyager 2 probes. *From NASA.*

We should not forget that the success of all missions flying beyond the earth and into space relies on our detailed understanding of the gravitational forces involved and our ability to use computers to discover the best possible trajectories.

7.5.2 Most Intriguing of All: No Escape

The universe is full of mysterious objects—quasars, pulsars, red giants, white dwarfs, dark matter, and so on. But in a popular vote for the most intriguing, I think the black hole might win. In fact, the idea of a "dark star" is an old one going back to the Reverend John Michell (1724–1793). (Laplace also put forward a similar idea in 1796.) Michell is little known today, but he was a geologist of some note. It was he who designed the torsion balance method for measuring gravitational effects, although today it is associated with Henry Cavendish who first performed the experiments.

Equation (7.6) lets us ask about the size and mass of a body for which the escape velocity is the speed of light c. Substituting $v_{esc} = c$, we find that

$$c = \sqrt{\frac{2GM}{R}} \quad \text{or} \quad \frac{M}{R} = \frac{c^2}{2G}, \quad\quad (7.7)$$

$$\text{or} \quad R = \frac{2GM}{c^2}.$$

This tells us that if a body is massive enough—or small and dense enough—the escape velocity may be as large as the speed of light. Here, in John Michell's own words, are the consequences:

> If the semi-diameter of a sphere of the same density with the sun were to exceed that of the sun by 500 to 1, a body falling from an infinite height towards it, would have acquired at its surface a greater velocity than that of light, and consequently, supposing light to be attracted by the same force in proportion to its vis inertiae, with other bodies, all light emitted from such a body would be made to return towards it, by its own proper gravity.[15]

Michell is telling us that for the right kind of massive body, no light will be able to entirely escape from it and so it will be dark; the escape velocity is greater than the speed of light.

The simple argument above is very clear, but it rests on Newton's idea that light consists of particles—which he called "corpuscles"—that behave like other bodies under the action of gravity. (More on this subject will come in chapter 9.) This theory is now not accepted, and we must turn to Einstein's general theory of relativity for a different approach. In Einstein's theory, gravity manifests itself by changing the structure of space-time around massive bodies. Then light follows the geodesics located in that structure. In turns out that for a spherically symmetric, static body, the differential element in space-time is given by the Schwarzschild metric:

$$(ds)^2 = B(r)c^2(dt)^2 - \frac{(dr)^2}{B(r)} - r^2(d\theta)^2 - r^2\sin^2(\varphi)(d\varphi)^2,$$

$$\text{where} \quad B(r) = 1 - \frac{2GM}{c^2 r}. \tag{7.8}$$

The space coordinates are the spherical polars (r, θ, φ). It is not necessary to appreciate these equations in detail, but note that $B(r)$ indicates how gravity (through the body mass M and the gravitational constant G) affects the intervals. In particular, things become very strange when B becomes zero, which happens at the Schwarzchild radius r_s:

$$B(r_s) = 1 - \frac{2GM}{c^2 r_s} = 0 \quad \text{when} \quad r_s = \frac{2GM}{c^2}. \tag{7.9}$$

The conditions for and behaviors of black holes are obtained using arguments based on these metric properties (see Roos, chapter 3, for example). Notice that the critical Schwarzchild radius r_s is just the same as R in equation (7.7) for Michell's dark star.

Karl Schwarzchild published his solution to Einstein's equations in 1916, and his calculation has had a major impact on cosmology. In 1939, Robert Oppenheimer and Harland Snyder discussed the behavior of collapsing neutron stars that become extremely dense. The importance and neglect of this work is summed up in Freeman Dyson's comment revealing a most unfortunate coincidence:

> The neglect of Oppenheimer's greatest contribution to science was mostly due to an accident of history. His paper with Snyder, establishing in four pages the physical reality of black holes, was published in the *Physical Review* on 1 September, 1939, the same day Adolph Hitler sent his armies into Poland and began World War II.[16]

Ironically, it is for his work at Los Alamos, building the atomic bomb (more on this is in chapter 11), that Oppenheimer is known today. In 1958, David Finkelstein identified the Schwarzchild radius as defining a boundary which allowed things to pass only one way and thus the building up of black holes was explained. There is now much evidence for the existence of black holes, which gained their name from John Wheeler in 1967.

The calculations dealing with escape from the effects of gravity—whether in Newton's framework and for space travel, or in Einstein's terms and for explaining strange objects in our universe—have been of enormous importance in mankind's travels and picture of the universe. It is essential to add **calculation 25, escaping gravity** to my list.

7.6 JUST TOO BIG

I expect some readers will be horrified by the work in cosmology not included in this chapter. I have tried to give a selection indicating some of the important advances and also those that are reasonably accepted and supported by observational evidence. Speculative theories involving string theory and multiverses are excluded. I did waiver about including something more on the big bang and the standard model; in particular, the calculations leading to the prediction of the cosmic microwave background radiation could have found a place. (Readers wanting a simple, general introduction might try Coles's book and Weinberg's *The First Three Minutes*.)

Chapter 8

ABOUT US

*in which we see calculations about us
and how our bodies work.*

The natural, or biological, world tends to be complex, and it is not as easy to give the types of definitive laws and their precise mathematical expressions, which characterize the inanimate world of the physical sciences. Nevertheless, calculations have played an important part in the development of biology, and a few of my favorite examples are given in this chapter. They are chosen to illustrate advances made using calculations in six areas: processes in our bodies, the variability of humans across a population, the way in which populations grow, common patterns in animal physiology, advances in genetics, and modern medical diagnostic tools. It is only a small selection, but the results were truly revolutionary for medicine, genetics, and medical diagnostics.

(I remind the reader that the very first calculation—**calculation 1, Malthus on population growth**—discussed as a leading example in chapter 1, really belongs in this chapter, under the third of these areas.)

8.1 THE CIRCULATION OF THE BLOOD

From the Renaissance onward there was an emergence of learning and investigation in the arts and the sciences. Although the knowledge of the ancient world, preserved and extended by the Arabs, was still taught and revered, there was a gradual acceptance of the need to carefully scrutinize it and sometimes replace it entirely. Copernicus's move to a sun-centered solar system in 1543 is a classic example. The work of William Harvey discussed in this section provides another example, one that was to prove to be both radical and central to the development of biology and medicine.

William Harvey (1578–1657) was educated at universities in Cambridge and Padua. In 1602, he was made doctor of medicine at Cambridge, and he went on to become an eminent physician, anatomist, and surgeon. He was physician to King Charles I for fifteen years. Harvey embodied the new approaches to learning, as shown by his statement that in his lectures on anatomy he professed "to learn and teach anatomy, not from books, but from dissections, not from the positions of the philosophers but from the fabric of nature."[1] His special interest was blood, and by overthrowing ancient viewpoints, he created our modern approach to the subject. Those criticizing the old teachings often found themselves under attack, especially as the Church had incorporated much of that ancient thinking into its own teachings and dogma. It is remarkable then that Harvey's friend, the philosopher Thomas Hobbes, could write that Harvey was "the only one I know who has overcome public odium and established a new doctrine during his own lifetime."[2]

The circulation of the blood through the body and its pumping by the heart are essential for life in animals to continue. It might seem strange then that the description of what was involved, given by Galen (about 129–200 CE), was accepted for over a thousand years, even though it is completely wrong. It becomes less strange when we appreciate two points. First, as mentioned earlier, the views of the ancients were revered, and few people were willing to challenge them. Second, it is extremely difficult to physically investigate blood flow and observe the beating heart—interfere with the system and death soon follows.

Blood in the body appears to come in two forms, one scarlet and one purple. We now know that they are the oxygenated and deoxygenated forms of the same blood. Galen believed that there were indeed two types of blood. One was generated by the liver and then transported in the veins to various parts of the body where it was consumed. The arterial system originated in the lungs and carried a vivified blood and life-giving spirit to the rest of the body. These ideas were used to devise medical treatments, such as bloodletting, for dealing with different diseases and illnesses.

8.1.1 Harvey and Blood Circulation

To understand blood flow in the body is to understand the role played by the heart, lungs, liver, veins, and arteries. Harvey investigated all of these organs in a long series of experiments, many of them involving animals and gruesome procedures. The result was his book *Exercitatio Anatomica de Motu Cordis et Sanguinis in Animalibus* or *An Anatomical Disquisition on the Motion of the Heart and Blood in Animals*, published in 1628. (See the bibliography where the details of Andrew Gregory's valuable history *Harvey's Heart* are also given.) Harvey discovered the blood circulation system as we know it today. This was a revolutionary achievement, made all the more remarkable by the fact that Harvey was unaware of the capillaries (since they are so small), which connect veins and arteries to complete the circuit.

Galen's theory that blood is produced by the liver had to be totally discredited, and Harvey does this in chapter 9 of his book. Here, in Harvey's words, is the way the system works:

> First, the blood is incessantly transmitted by the action of the heart from the vena cava to the arteries in such quantity that it cannot be supplied from the ingesta, and in such wise that the whole mass must very quickly pass through the organ; second, the blood under the influence of the arterial pulse enters and is impelled in a continuous, equable, and incessant stream through every part and member of the body, in much larger quantity than were sufficient for nutrition, or than the whole mass of fluids could supply; third; the veins in like manner return this blood incessantly to the heart from all parts and members of the body.[3]

The concept of circulation, rather than continuous blood production by the liver, is central to Harvey's theory, and he needs all available evidence to convince people to give up the ancient and respected description of this vital bodily function. It is here that his calculation is introduced. Again, we can follow Harvey's own words from his chapter 9:

> Let us assume, either arbitrarily or from experiment, the quantity of blood which the left ventricle of the heart will contain when distended

to be, say two ounces, three ounces, one ounce and a half—in the dead body I have found it to hold upwards of two ounces. Let us assume further how much less the heart will hold in the contracted than in the dilated state; and how much blood it will project into the aorta upon each contraction—and all the world allows that with the systole something is always projected, a necessary consequence demonstrated in the third chapter, and obvious from the structure of the valves; and let us suppose as approaching the truth that the fourth, or fifth, or sixth, or even but the eighth part of its charge is thrown into the artery at each contraction; this would give either half an ounce, or three drachms, or one drachm of blood as propelled by the heart at each pulse into the aorta; which quantity by reason of the valves at the root of the vessel, can by no means return into the ventricle. Now in the course of half an hour, the heart will have made more than one thousand beats, in some as many as two, three, and even four thousand. Multiplying the number of drachms propelled by the number of pulses, we will have either one thousand half ounces, or one thousand times three drachms, or a like quantity of blood, according to the amount which we assume as propelled with each stroke of the heart, sent from this organ into the artery, a larger quantity in every case than is contained in the whole body![4]

The calculation is simple: the amount of blood flowing as sent out by the heart in a certain time is given by the amount at each contraction multiplied by the number of heartbeats in that time. The exact details are not important; all of the assumptions show that there is more blood in this flow than could possibly be made by the liver. Galen's revered theory is discredited, and a revolution has begun in medicine. **Calculation 26, Harvey establishes blood circulation** is almost trivially simple, but its impact means that it must be on my list of important calculations. It helped to overturn over a thousand years of medical practice and ushered in our present-day understanding of how animal bodies function.

8.1.2 The Scientific Setting

The scientific revolution relied on three important approaches in science: the use of experiments (rather than relying on old written accounts), the use of a mechanical description and discrete or atomic theories of matter, and the use of mathematics. The work of Galileo, Boyle, Newton, and their contemporaries is often taken as instigating the scientific revolution. It might seem that Harvey was the great pioneer in medical science and biology, however, the real story is more involved (see Gregory).

Certainly, Harvey says the following of "true philosophers":

> nor are they so narrow-minded as to imagine any of the arts or sciences transmitted to us by the ancients, in such a state of forwardness or completeness, that nothing is left for the ingenuity and industry of others.[5]

But Harvey does display great respect for those ancients, referring to the "divine Galen" and making many references to Aristotle. The calculation, which to us seems so conclusive, is a tiny part of Harvey's book. In the final paragraph, he mentions the many supporting points "in the course of dissections" that confirm his theory, rightly emphasizing the importance of the great range of experiments he conducted, but with no mention of calculations. One of Harvey's critics was the German professor Caspar Hoffman, and, according to Andrew Gregory,

> Hoffman accused Harvey of "abandoning anatomy for logistics," in making calculations about the heart and blood flow, and said that: "Truly, Harvey, you are pursuing the incalculable, the inexplicable, the unknowable."[6]

This comment seems a little unfair. Harvey clearly did respect the ancients, but his work surely inspired those who saw that it was time to challenge and overthrow some of the old ideas and prejudices. The use of a simple but irrefutable calculation was a powerful factor in that challenge.

8.2 COLLECTING AND ANALYZING DATA
ON LIFE AND DEATH

In the previous section, we saw how Harvey discovered the blood flow mechanism operating in all animals. Although the bodily function mechanisms might be the same in all cases, no two humans are exactly the same, and the variations are of great interest. Biology is based on samples of organisms, and collections of data allow us to identify general trends, study the range of variations, and draw some general conclusions. The work discussed in this section is some of the earliest of this type relating to human populations. Another example involving a spread of species will be given later.

We are all born and we all die. But how long we live and how we die vary enormously. The first in-depth study of these variations is often taken to be the work of John Graunt (1620–1674). Although he was a draper, Graunt developed an interest in the London Bills of Mortality and the information that might be extracted from them. The Bills of Mortality listed the number of people dying from certain causes over a given week and also gave information about christenings, funerals, and other events (an example is given by Lewin and De Valois). Graunt's very well-received small book, *Natural and Political Observations Made upon the Bills of Mortality*, was published in 1662, and it led Charles II to have Graunt made a member of the Royal Society—quite an achievement for a "haberdasher of small-wares."[7] Graunt tried to estimate both the population and its breakdown (males/females, married/single, and different age groups). Of course, one aim of the bills was to provide information about the occurrence of the plague and movements of people.

The data in the Bills of Mortality is interesting if you want to know how many people died from "Griping in the Guts" or how many people were "Murthered" (in fact, surprisingly few), but it was not easy to extract precise data on population trends. Some things called for an explanation; in his conclusion, Graunt said he wished to know "why the Burials in London exceed the Christenings, when the contrary is visible in the Country."[8] London was a city of comings and goings, and the population did not have the underlying stability required for demographic studies.

8.2.1 The Breslau Data Becomes Available

The registers of births and deaths were very detailed and carefully kept in the city of Breslau in Silesia. Evangelical pastor Caspar Neumann (1648–1715) used them to fight popular superstitions such as the idea that health depended on the phases of the moon. Neumann passed his data on to Gottfried Wilhelm von Leibniz, from whom it went to France and then on to London where its value was finally recognized. Most importantly, the data for the years 1687–1691 were relatively stable; the city of Breslau did not have the continuous influx of people that so confused the data for London. The person who recognized the true value of the Breslau data was Edmond Halley—yes, he of the comet and transit of Venus fame!

In1693, Halley published a groundbreaking paper, "An Estimate of the Degrees of the Mortality of Mankind, Drawn from curious Tables of the Births and Funerals at the City of Breslaw," in the Royal Society's *Philosophical Transactions*. (Excerpts from both Graunt's book and Halley's paper are available in Newman's *The World of Mathematics*.) In this paper, Halley demonstrated how to organize data into a useful table, how to make immediate observations and deductions, and how to make use of this data in calculations. After suitably organizing the data (rounding the population aged under one to 1,000, for example) Halley produced the table shown in figure 8.1. This table shows how the population of 34,000 was distributed over the different age levels, both year by year and in groups of seven years.

Having organized the data into a suitable form, Halley could extract information from his table "whose uses are manifold."[9] His first use was to find "the proportion of men able to bear arms." (Graunt tried a similar calculation for London.) After defining this group as men aged between 18 and 56, it was a simple matter of adding up the numbers in the table and assuming roughly half were males. His result was "about 9000, or 9/34, or somewhat more than a quarter of the Number of Souls, which may perhaps pass for a Rule for all other places." Halley is making an important point: deductions using the Breslau data might be valid for other populations.

Age. Curt.	Per-fons.	Age. Curt.	Per-fons	Age. Curt.	Per-fons	Age. Curt.	Per-fons	Age. Curt.	Per-fons	Age. Curt.	Per-fons
1	1000	8	680	15	628	22	586	29	539	36	481
2	855	9	670	16	622	23	579	30	531	37	472
3	798	10	651	17	616	24	573	31	523	38	463
4	760	11	653	18	610	25	567	32	515	39	454
5	732	12	646	19	604	26	560	33	507	40	445
6	710	13	640	20	598	27	553	34	499	41	436
7	692	14	634	21	592	28	546	35	490	42	427

Age Curt	Per-fons.	Age. Curt.	Per-fons	Age. Curt.	Per-fons	ge Curt	Per-fons	Age. Curt.	Per-fons	Age. Crfr.	Per-fons
43	417	50	346	57	272	64	202	71	131	78	58
44	407	51	335	58	262	65	192	72	120	79	49
45	397	52	324	59	252	66	182	73	109	80	41
46	387	53	313	60	242	67	172	74	98	81	34
47	377	54	302	61	232	68	162	75	88	82	28
48	367	55	292	62	222	69	152	76	78	83	23
49	357	56	282	63	212	70	142	77	68	84	20

Age.	Perfons.
7	5547
14	4584
21	4270
28	3964
35	3604
42	3178
49	2709
56	2194
63	1694
70	1204
77	692
84	253
100	107
	34000
	Sum Total.

Figure 8.1. Halley's table of population distribution in Breslau. *From Edmond Halley, "An Estimate of the Degrees of the Mortality of Mankind, Drawn from Curious Tables of the Births and Funerals at the City of Breslaw, with an Attempt to Ascertain the Price of Annuities upon Lives," Philosophical Transactions 17 (1693).*

Halley's second use was to show the "Vitality in all Ages," by which he means the chances of someone at one age living to be another. For example, he shows that the chance of a person age 25 not dying before turning 26 is 560 to 7, or 80 to 1, because his tables show that the 567 persons alive at 25 have reduced to 560 at age 26. Similarly he shows that "it is 377 to 68, or 5½ to 1, that a man of 40 does live 7 years." Similarly, in his third use, Halley shows how to calculate the odds of someone dying or life expectancy. In his example, he shows that "a man of 30 may reasonably expect to live between 27 and 28 years."

8.2.2 The Mathematics of Insurance

Halley's first uses of his Breslau table simply calculate numbers revealing how the population develops and ages. In the next uses, Halley turns to how the results may be applied, and in doing so, he begins a branch of mathematics that has been of commercial importance ever since. Halley makes clear his intentions for his fourth use, which is worth quoting in full:

> By what has been said, the Price of Insurance upon Lives ought to be regulated, and the difference is discovered between the price of ensuring the life of a man of 20 and 50, for example: it being 100 to 1 that a man of 20 dies not in a year, and but 38 to 1 for a man of 50 years of age.[10]

His fifth use begins: "On this depends the Valuation of Annuities," and it is here that Halley establishes what is often called the cornerstone of actuarial science. (In fact the full title of Halley's paper continues from that given above with the words *with an Attempt to ascertain the Price of Annuities upon Lives*.) There has always been a desire and a need to arrange funds for a secure and comfortable life in old age. One way is to purchase an annuity, whereby a sum of money is paid by a person to a government or other organization so that later a steady income is guaranteed after some defined year (often the year of the person's retirement) and usually until death. It sounds like a dry and boring topic, but annuities go back to at least Roman times. (Their surprising history can be found in the works of Ciecka, Hald, Kopf, and Lewin and De Valois.)

The basic question is: How much should one pay for an annuity? Halley has been considering the chances of people living beyond any given age, and he now wants this to be used in answering that question:

> for it is plain that the purchaser ought to pay for only such a part of the value of the Annuity, as he has chances that he is living; and this ought to be computed yearly, and the sum of all those yearly values being added together, will amount to the value of the annuity for the Life of the Person proposed.

There is also the point that the organization providing the annuity will invest the funds paid to generate the annuity at a rate of interest p, and this is compounded to add to the deposited amount. Thus the money available to the provider is actually the sum paid plus the interest to be earned on that sum. Obviously the person buying the annuity should be aware that the sum paid need not be as big as at first expected when those extra earned interest contributions are taken into account. Halley showed how to evaluate the cost of annuities when these points are properly considered. He used the example of interest at 6 percent, so $p = 0.06$.

Halley explains his method in prose rather than in mathematical notation making it hard to follow. It amounts to using the following formula (see Ciecka or Hald) for the cost of an annuity paying unit amount per year after year n:

$$\text{cost of annuity} \quad = \quad \sum_{m=1}^{death} (1+p)^{-m} \left(\frac{L_{n+m}}{L_n} \right).$$

The quantities L_j (used in the above equation with j equal to n and $n + m$) are given by the entries in Halley's table (figure 8.1) for those living in year j, and their ratios give the chances of surviving into future years. The upper limit on the sum means carry on until there are no survivors for that year. Unless a computer is used, "this will without doubt appear to be a most laborious calculation," as Halley puts it. However, "after a not ordinary number of Arithmetical Operations" he produces a table of annuity values for five-year intervals for ages 1 to 70.

In uses six and seven, Halley considers the cases where more than one life is involved. This is a complicated problem, but he is still able to use his stated calculating principles.

Selling annuities was a way for governments to raise money, and this was happening in England during Halley's time. Clearly the government had no idea about the mathematics involved or how to properly cost an annuity; while they were instantly gaining money (to use for running the country), they were going to lose out when the time came to pay the annuities. Halley saw the opportunity; using the results in his table, he says, "this shows the great advantage of putting money into the present fund lately granted to their Majesties." I am not sure when a government official finally caught up with the theory and realized they were running an over-generous insurance scheme!

Edmond Halley may be best known for his comet, but his clever and innovative pioneering work in actuarial science makes **calculation 27, Halley values annuities**, a worthy addition to my list of important calculations. (Work immediately following on from Halley's is described by Hald, and Lewin and De Valois. The use of information from large databases and extensive surveys is now a standard part of the biological and social sciences; see Cohen for a gentle introduction.)

8.3 AN EARLY FUNDAMENTAL ADVANCE IN GENETICS

The nineteenth century saw the development of Mendelian genetics and Darwin's theory of evolution by natural selection. In the twentieth century, they came together to give our modern framework for biology, and the mathematical basis was built up by R. A. Fisher, J. B. S. Haldane, and Sewall Wright. The result was the subject of population genetics, and it is the derivation of a first significant result in this field that is the subject for this section. (An easily accessible and readable introduction is given by Samir Okasha.) The result incorporates Mendel's discoveries and tackles one of Darwin's worries: How can variability be maintained in breeding populations that he saw as blending the characteristics of the individuals?

8.3.1 Some Background

We consider a large population of sexually reproducing organisms. The organisms are taken to be diploids: each cell contains two copies of each chromosome, one inherited from each parent. In order to reproduce, the parents produce gametes (which fuse together in sexual reproduction), and these gametes are haploid: they contain only one of each chromosome pair. The fusion process gives a new cell or zygote, which is again diploid, and this gives rise to the new organism. This is the life cycle followed by most multicelled animals and many plants.

The most basic problem arises when one particular chromosome locus or slot (or gene as it has now become), has two possible forms, known as alleles, which I denote by A and a. In the organism or phenotype, they will be responsible for some particular characteristic. For example, it could be the gene controlling eye color, with A giving brown eyes and a giving blue eyes. Or it could be the smooth or wrinkled seeds, the large or small plant heights, or the white or purple flower colors, in Mendel's pea-breeding experiments.

Over a large population, the alleles will occur with particular frequencies $f(A)$ and $f(a)$. (You may think of the fractions of each found in a large survey.) I will let

$$f(A) = p \quad f(a) = q \quad p + q = 1. \tag{8.1}$$

For example, if 80 percent of the time we find the A allele and 20 percent the a allele,

$$f(A) = 0.8 \quad f(a) = 0.2 \quad 0.8 + 0.2 = 1. \tag{8.2}$$

Clearly those will also be the frequencies of the alleles occurring in the gametes produced by the parents for sexual fusion. However, the zygotes so produced will have a pair of chromosomes with the particular chromosome locus (gene) under consideration being either A or a according to the gametes used to produce them. Thus we can now work out the frequencies of finding the various pairs of alleles—$f(AA)$, $f(Aa)$, and $f(aa)$—in the zygotes (and hence in the new organism as it develops after the sexual reproduction). Using the frequencies for the alleles themselves in the gametes we are led to

$$f(AA) = p^2 \quad f(Aa) = 2pq \quad f(aa) = q^2. \tag{8.3}$$

The frequencies should add up to one, which is correct if we note equation (8.1):

$$p^2 + 2pq + q^2 = (p + q)^2 = 1^2 = 1.$$

For the numerical example in equation (8.2),

$$f(AA) = 0.64 \quad f(Aa) = 0.32 \quad f(aa) = 0.04 \quad \text{and} \quad 0.64 + 0.32 + 0.04 = 1.$$

We now know how the alleles are distributed in the new generation of organisms. For example, if we check the pair of chromosomes having the slot or locus under discussion, we will find one has the A allele and the other the a allele in 32 percent of the cases. They both have the a allele in 4 percent of the cases.

8.3.2 Analysis and Concerns

Suppose that when the genes are expressed, allele A is dominant and a is recessive. For example, brown eyes must follow when we find the AA or Aa pairs, but blue eyes only follow from the aa pair. Thus in the new generation, for our numerical example, we expect to find 4 percent of individuals with blue eyes. Going back to the parent population, the allele for blue eyes actually occurred in 20 percent of cases according to equation (8.2). It appears that the dominant allele has led to a drastic reduction in blue eyes and the natural feeling is that if the process repeats to give the next generation, blue eyes will be even further reduced in number. The apparent conclusion is that the dominant form of the gene will quickly rule the whole population.

This was a major problem for biologists early in the twentieth century: How is genetic variation maintained in a sexually reproducing population?

The solution to the problem involves a little careful analysis of the breeding process. The zygotes in the population with parents having allele frequencies as in equation (8.1) were shown to have allele pairs with frequencies according to equation (8.3). Now we ask: what will be the allele frequencies that we must use when this new, first generation becomes parents and produces the second generation? We can find that out by counting how the alleles occur in that new first generation and that is given by equation (8.3). The A allele occurs whenever the AA pair occurs, and in half the cases when the Aa pair occurs. Thus we find that the single chromosome put into the gamete has the A allele frequency

$$f_1(A) = p^2 + pq = p(p + q) = p \times 1 = p. \qquad (8.4a)$$

Similarly, we find the a allele frequency in the new first generation to be

$$f_1(a) = q^2 + pq = q(p + q) = q \times 1 = q. \qquad (8.4b)$$

Now amazingly, comparing equations (8.1) and (8.4), we see that the first new generation actually has allele frequencies f_1 which are the same as those f for their parents.

We now have the essential result: breeding has not changed the allele frequencies. So, when the new generation breeds to give the second generation, the same result must occur and the distribution of allele pairs in the new zygotes will again be as in equation (8.3).

The dominant allele does not swamp out the genetic variation; in this way of reproducing, the genetic variation is stable. This result is known as the Hardy-Weinberg law, which was announced in 1908. It tells us that the blending problem worrying Darwin does not exist, and using Mendel's more particle-like concepts, the variation in a population is maintained.

The message from the Hardy-Weinberg law is so significant that it is worth emphasizing: there is equilibrium in the genetic makeup of a sexually reproducing population of organisms so that variability is maintained. There are other ways to analyze the reproducing system described above, but they all come down to some sort of counting process and are basically simple calculations despite the attention to detail needed to follow them. This led the geneticist W. J. Ewens to conclude that "it does not often happen that the most important theorem in any subject is the easiest and most readily derived theorem for that subject."[11] You will hardly find it surprising that I am adding **calculation 28, the Hardy-Weinberg law** to my list of important calculations.

8.3.3 Introducing Variability

You may be saying: "but there really is variability, otherwise there would be no striking changes in populations and evolution." The solution to this dilemma is to note that in deriving the Hardy-Weinberg law several assumptions have been made:

- Organisms are diploid and only sexual reproduction occurs.
- Mating is at random.
- Generations are nonoverlapping.
- Genders are evenly distributed among the three genotypes and all genotypes are equally fit.
- The population is very large (theoretically infinite).
- There is neither immigration or emigration.
- There is no mutation or artificial selection.

If any of these conditions are violated, the Hardy-Weinberg law does not strictly hold in the form given above. In fact, checking the allele frequencies and genotypes is a way to check whether a population is in Hardy-Weinberg equilibrium and so to detect the influence of one of these conditions. Of course, the conditions will be violated in some ways and to certain degrees; mutations will occur, and hence there will be evolution as Darwin envisaged it.

The analysis can be extended in various ways and new results produced. For example, the consequences of overlapping generations and multiple genes and alleles must be considered. There is now an enormous body of work in genetics and its mathematical treatment.

8.3.4 The Remarkable Discoverer

Hardy and Weinberg discovered their law in 1908. Wilhelm Weinberg was a German physician, and his work was not well known in the English- speaking world until 1943. For some time, the result was known as Hardy's law, and it was given in Hardy's 1908 *Science* paper. G. H. Hardy (1877–1947) was responding to claims that brachydactyly (a defect giving short fingers), through its genetic dominance, should be widespread and eventually afflict three quarters of the population. He constructed an argument using alleles A and a with frequencies p and q which led to the result given above He found what he termed "the stable distribution" and showed that the worries about widespread brachydactyly were unfounded since its occurrence is rare and will remain so whether the gene for it is dominant or recessive.

The intriguing part of this story is that Hardy was one of the leading pure mathematicians of the first half of the twentieth century. His book on number theory remains a classic. In his final, wonderful book about his life and mathematics, *A Mathematician's Apology*, he expressed his disdain for applied mathematics and wrote that "I have never done anything 'useful'. No discovery of mine has made, or is likely to make, directly or indirectly, for good or ill, the least difference to the amenity of the world."[12]

According to Hardy, the mathematics used in applications tends to be less demanding and innovative than that explored by the pure mathematician, and in his *Science* paper he writes of "the very simple point" that he wishes to make. Of course, it is often a simple mathematical argument that leads to a profound result in science, as several of the calculations discussed in this book illustrate, but that was not an appealing point for Hardy. It is ironic then, that Hardy the pure mathematician is probably little known compared with the Hardy in the Hardy-Weinberg law, which plays such an important and fundamental part in genetics.

8.4 THE MATHEMATICS BEHIND COMPUTED TOMOGRAPHY (CT)

In 1895, Wilhelm Röntgen discovered x-rays, which were soon used for medical diagnosis. For the first time, doctors could see inside the body and detect things like broken bones; the medical scene changed dramatically. However, the common x-ray gives only an overall picture produced by the total x-ray absorption profile of the whole body part through which the x-rays pass, and no details of the individual components involved are given. Early in the twentieth century, people began to think about "body-section radiography." In 1937, Edward Wing Twining (1887–1939), the father of British neuroradiology, introduced the term "tomography"—from the Greek, tomos for slice, and graphein, to write. (For a history of tomography see the book by Webb and the article by Gordon, Herman, and Johnson.) The problem concerns how to use several x-rays transmitted through a section of a body or body part in order to deduce the details of the internal makeup of that section. This is an inverse problem much like that faced by seismologists using seismic waves to discover what is deep inside the earth, as explained in section 4.4. Perhaps the best way to explain the situation and how it may be tackled is to introduce a small, or "toy," example.

X-rays are partially absorbed as they pass through the body with different materials—muscles and bones, for example—absorbing different amounts. It is this differential absorption rate that gives the variable x-ray beam that is then used to create a picture on the x-ray plate. The problem is how to use these different absorption rates to build up a detailed picture of the inside of the body.

8.4.1 A Simple Example Shows the Way

Consider a box or grid split into nine equal parts as shown in figure 8.2. This could represent a square section of material with x-ray absorption coefficients as indicated—the top corner part has absorption set at unity, and the other parts have unknown absorptions denoted by x_i. If x-rays are scanned through the section, can we determine those x_i?

Figure 8.2 indicates how eight scans may be made and gives some possible numerical outputs for measured absorptions. Thus the first scan tells us that $x_6 + x_7 + x_8$ is equal to 13, but it gives no information about their individual values. They could be 3, 4, and 6, or 2, 9, and 2, for example. The same thing happens with each of the other seven scans; we get information about sums of absorptions, but not individual values.

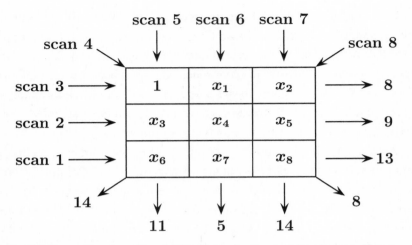

Figure 8.2. Scanning along the eight lines to give the indicated sums. *Figure created by Annabelle Boag.*

Notice that if the individual absorptions were given, it is a trivial matter to calculate the result of any scan by a simple addition. The important question is: Can we solve the inverse problem—can we use all the given scanning information to extract the individual absorption values? If we write out the measured information in mathematical form we get

$$
\begin{aligned}
&\text{scan 1:} &&& x_6 + x_7 + x_8 &= 13 \\
&\text{scan 2:} && x_3 + x_4 + x_5 && &= 9 \\
&\text{scan 3:} & 1 + x_1 + x_2 &&& &= 8 \\
&\text{scan 4:} & 1 \quad\ + x_4 &\quad\ + x_8 &&= 8 \\
&\text{scan 5:} & 1 \quad + x_3 &\quad + x_6 &&= 11 \\
&\text{scan 6:} & x_1 \quad + x_4 &\quad + x_7 &&= 5 \\
&\text{scan 7:} & x_2 \quad + x_5 &\quad + x_8 &&= 14 \\
&\text{scan 8:} & x_2 + x_4 + x_6 &&& &= 14
\end{aligned}
$$

This is a set of eight linear equations for the eight unknowns, x_1, x_2, . . . x_8, and there are standard, straightforward ways to solve them to get the values 2, 5, 3, 2, 4, 7, 1, and 5.

This example illustrates the basic point that by employing a mathematical method, it is possible to use the data from several individual scans to calculate the internal structure of the sample being examined. An important note: while it is mathematics that allows the inverse problem to be solved, in practice, a computer is required to carry out the detailed calculations suggested by the mathematics. The set of equations given above could be solved without using a computer, but it is a tedious and time-consuming business.

Notice that the practical requirement has changed because we now need a machine to record the numerical values of the absorptions; the output of the x-ray is now a numerical data set rather than a picture on a film.

The processing of the data raises many questions related to accuracy and the influence of experimental error on the final output. There is also a fundamental question: Does the algorithm produced by the mathematical analysis of the general problem lead to a unique answer? In effect, can we really get the correct picture of what is inside the body section being examined? The solution to the set of linear equations used in the above example is known to be unique. It is possible to extend that method to produce a viable tomography technique, although there are problems in the detailed applications (see the article by Gordon, Herman, and Johnson, or chapter 7 in the book by Kak and Slaney).

A little more of the relevant mathematics is given in the next section, but readers not interested in such matters may skip on to the actual historical development of tomography.

8.4.2 The General Problem

If an x-ray beam of initial intensity I_0 travels along a line L through a material with absorption coefficient $g(s)$, where s is distance along the line, then the emerging beam has intensity

$$I = I_0 \exp\left\{-\int_L g(s)\,ds\right\}. \tag{8.5}$$

If we define f_L by

$$f_L = \int_L g(s)\, ds \qquad (8.6)$$

we have

$$I = I_0 \exp\{-f_L\} \quad \text{which gives} \quad f_L = \log_e (I_0/I).$$

Thus the values of I give the values of f_L. Does this lead to $g(s)$, which is what we really want?

Mathematically, we have the following problem: suppose that g exists everywhere for a finite region in a plane, and we may form the integrals along all possible lines L passing through that region to give all f_L as in equation (8.6). The question is whether or not we can use all the f_L data to find the values of g at all points in the region of interest.

The answer to this question was actually given in 1917 by the Austrian mathematician Johann Radon, and the process of going from g to the set of line integrals f_L is known as taking the Radon transform. Radon showed that the inversion problem could be uniquely solved; that is, given the line data, we can construct the function g over the region of interest. Thus the central mathematical question for tomography had already been answered in 1917, but most of the people working in the x-ray field half a century later were unaware of Radon's work and made the analysis again for themselves.

It is important to understand that knowing that a problem has a solution in theory does not mean it will be simple to take it over for practical applications. In this case, there needs to be an enormous amount of work done to establish a viable inverse procedure and to convert the whole formalism to one in which discrete, rather than continuous, data sets may be used. Obviously, in practice, we cannot scan along all possible lines, although with the latest technology a very large number can be used. (There is now a large literature on this subject, and in the bibliography, I have given references to an early mathematical discussion by Shepp and Kruskal, and the books by Kak and Slaney, and Epstein.)

8.4.3 Cormack's Contribution

Allan Cormack published his paper "Representation of a Function by Its Line Integrals, with Some Radiological Applications" in 1963, and he published a second part of his work in 1964. In essence, Cormack set up the problem as explained above and then showed how it could be completely solved using a Fourier-series method and numerical analysis. The uniqueness of the solution was also examined.

Cormack went further and proved the viability of his method using a disc specimen (often referred to as a phantom) and gamma rays rather than x-rays. His first specimen had circular symmetry as shown in figure 8.3 (a). It comprised a section of an aluminum cylinder of diameter 10 cm with a wooden surround of outer diameter 20 cm. The symmetry meant that lines at only one angle needed to be sampled, and Cormack scanned the beam as shown in figure 8.3 (b) with a line spacing of 5mm. The results given by Cormack's calculations proved to be accurate, so much so that the absorption coefficient of the aluminum section was shown to be slightly different in an inner and outer part. It turned out that when the specimen was manufactured, an inner part used pure aluminum, but then an alloy was used to complete the 10 cm diameter disc. In the next stage of his work, Cormack used a phantom lacking circular symmetry for which lines at different angles had to be scanned, see figure 8.3 (c) and (d). Again, he obtained impressive results.

Allan Cormack received the Nobel Prize for this work, and his highly informative acceptance address reviewing his work is given in his 1980 *Science* article. He remarks that his work did not gain immediate recognition, but he did get a request from the Swiss Center for Avalanche Work asking about the relevance to objects buried in snow!

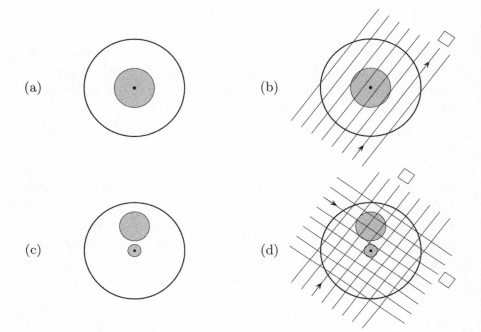

Figure 8.3. A sketch of Cormack's scanning experiments. (a) and (c) show the type of specimens used, with shading to indicate the aluminum region. (b) and (d) indicate the scanning lines used. In the second case, scanning lines at twenty-five angles were used. *Figure created by Annabelle Boag.*

8.4.4 Hounsfield's Contribution

The Nobel Prize was shared by Godfrey Hounsfield (or Sir Godfrey, as he became in 1981). Hounsfield worked for EMI in England and produced the first viable scanners for medical purposes. His first patent was granted in 1972. An idea of the difficulties involved in operating the first CT scanners can be gained from the fact that an early test involved the recording of 28,000 measurements on paper tape; the data took nine days to collect, and the available computer took two and a half hours to analyze it. In one of the first clinical applications, a frontal-lobe tumor was detected in a 41-year-old patient. (See Webb for details and references.) Since then, CT scanning has gone from strength to strength as more sophisticated scanning machines and computers have been developed. Today, CT scans have revolutionized many areas of diagnostic medicine. (As an amusing aside, it is claimed that EMI could sponsor Hounsfield's research because it made large sums of money selling the Beatles' records!)

8.4.5 Should It Be CMT Scan?

The invention of the CT scanner was a major step in medical science. Those early pioneering efforts by people like Cormack and Hounsfield have given rise to a wonderful tool making life easier for millions of people every day. The mathematics which is such an essential part of computed tomography is often forgotten, or certainly not fully appreciated. Without the mathematical basis there would be no CT scanner; equally, we must recognize that without significant computer power there would be no way to use that mathematical basis. Personally I would like to see this great tool referred to as "computed mathematical tomography" or CMT. It is probably too late for such a renaming, but at least I can add **calculation 29, the mathematics behind the CT scan** to my list of important calculations.

8.5 LINKING ALL THE ANIMALS

Darwin's ideas explain how different species can evolve and give us the great diversity of life that confronts us today. But this same theory also tells us that all life is somehow linked, and today we marvel at the similarities in the DNA of quite different creatures. Harvey showed how the blood of all animals is circulated with the same mechanism involving a heart, veins, and arteries. Similarly, there is a nervous system carrying messages around inside animal bodies. These are general properties, and we know they follow from nature's blueprint for the growth and reproduction of life. But does this blueprint impose regularity beyond the mere similarity of the processes involved and the apparatus that grows to facilitate them? Can we find general principles that tell us how to describe these regularities and then how to explain them? This section is concerned with one approach to that question, and it relies on simple but powerful ideas about the scaling of physical objects. As an example, I ask what can be said about vertebrate animals as their size varies. As a measure of size, I take the mass M of the animal, which is the most simply defined and easiest quantity to measure consistently.

8.5.1 The Basic Mathematics of Scaling

We are used to seeing children's toys, such as cars or airplanes, as scaled-down versions of the real things, but probably few people naturally think of their pet dog as a scaled-up version of a mouse, or a scaled-down version of a horse or even an elephant. Does such a scaling idea make sense? Is it a useful idea? To explore these questions it is necessary to find a mathematical basis for the scaling concept.

Consider a cube of material of side length L and density ρ. We can then find the following relationships:

$$\text{length } L, \qquad \text{surface area } A = 6L^2,$$
$$\text{volume } V = L^3, \qquad \text{mass } M = \rho V = \rho L^3. \tag{8.7}$$

I have used the length L as the basic quantity, and the mass is proportional to its cube. I can also use the mass M as the basic quantity, and the volume is proportional to the mass. These ideas carry over to shapes other than cubes. For all quantities, I can write:

$$V \propto M \qquad A \propto M^{2/3} \qquad L \propto M^{1/3}. \tag{8.8}$$

In general, for any quantity y, I write:

$$y \propto M^b \quad \text{or} \quad y = aM^b. \tag{8.9}$$

Equation (8.9) describes how some quantity y scales as the mass varies; b is called the scaling exponent, and a is the proportionality constant. For example, if we put y equal to the area, we use $b = \tfrac{2}{3}$.

Equation (8.9) is used to investigate the scaling properties of animals, and y might be, for example, lifespan, lung volume, brain size, or skeleton size.

There is an important mathematical step that is taken to make the scaling easier to handle and interpret. We make use of logarithms (to base 10 here) as introduced in chapter 3. If we take the log of equation (8.9) we get

$$\log(y) = \log(a) + b \log(M). \qquad (8.10)$$

If we plot the logarithm of y against the logarithm of M, equation (8.10) tells us that we will get a straight line with slope b. Straight lines are very easy to deal with, and the comparison curves are illustrated in figure 8.4. Notice also that if M varies from 1 to 10 to 100 to 1,000, the log (to base ten) varies from 0 to 1 to 2 to 3, meaning that the enormous range from 1 to 1,000 is simply 0 to 3 on the log scale. This is usually called plotting on a log-log scale. Notice also that it is very easy to read off the slope of a line and hence find the exponent b.

(Readers wishing to see a more expansive introduction to the subject should seek out the classic *Scaling: Why Is Animal Size So Important?* by Knut Schmidt-Nielsen, or introductory biology books such as those by Burton, Schmidt-Nielsen (1972), and Vogel.)

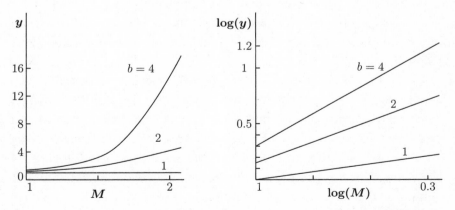

Figure 8.4. Examples of how curves of the type in equation (8.9) translate into straight lines when plotted on a log-log scale. *Figure created by Annabelle Boag.*

8.5.2 A First Example

The calculations we are considering take data about some quantity and investigate whether it follows a scaling law of the form described by equations (8.9) and (8.10). As an example, I take the mass of mammalian skeletons as studied by H. D. Prange and others (see also Schmidt-Nielsen, chapter 5). They took the data, converted it to logarithmic form, and produced graphs like the simplified version shown in figure 8.5.

The first thing we gain from figure 8.5 is that there does appear to be a relationship between the mass of the skeletons of animals and their total mass. Of course, there is some scatter, as there always will be in biological data. However, the trend, or regularity, is very clear; remarkably, it covers around five orders of magnitude in the masses involved, from animals weighing a few grams to those weighing hundreds of kilograms.

As a first guess, we might expect the skeleton mass M_{skel} to increase in exactly the same way as the total mass, but this is not what is revealed in figure 8.5:

$$\text{expected } M_{skel} = aM \qquad \text{according to the data } M_{skel} = aM^{1.09}.$$

According to the calculation made by Prange and his colleagues, the scaling exponent is 1.09 rather than 1; the graph in figure 8.5 does not have slope equal to one. This implies that if the mass M increases by a factor of ten, we expect that the mass of the skeleton increases according to the factor $10^{1.09} = 12.3$ times.

This is an example of the importance of the exponent in scaling calculations. We now need to ask why the exponent takes on its particular value. I have chosen this case because it is easy to think about bone sizes or cross-sections, the weight to be supported by the bones without them being crushed, and the different lifestyles of the animals involved. Figure 8.6 shows the skeletons of a cat and an elephant drawn to the same size. It is quite clear that the cat is much finer-boned; we know that it has relatively little weight to support, and it readily jumps and prances around in ways that an elephant does not. Interestingly, Galileo was already writing about these variations in bone sizes in his 1638 *Two New Sciences*. We should expect heavier animals to have more massive skeletons to support them, and we can argue for scaling exponents greater than one.

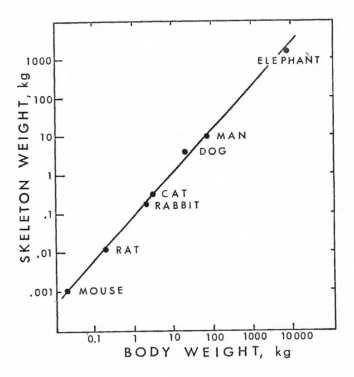

Figure 8.5. Skeletal mass versus body mass for mammals plotted on a log-log scale. *Reprinted with permission, © Cambridge University Press, from K. Schmidt-Nielsen,* How Animals Work *(Cambridge: Cambridge University Press, 1972).*

This example illustrates four values or roles of scaling calculations:

1. Organizing data
2. Exhibiting trends and regularities
3. Providing a parameter to precisely characterize those regularities
4. Suggesting the theories that must be developed to understand the origin of scaling exponents

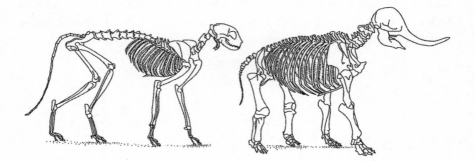

Figure 8.6. Skeletons of a cat (*left*) and an elephant (*right*) drawn to the same scale. *Reprinted with permission of Princeton University Press, from Steve Vogel,* Life's Devices: The Physical World of Animals and Plants *(Princeton: Princeton University Press, 1989).*

8.5.3 Regularity in the Fire of Life

Animals use the same chemical processes to maintain life, and surely the most remarkable (and certainly the most debated) scaling result concerns the variations in how these processes operate as a function of the size of the animals involved. Animals use chemical reactions to produce energy and a variety of products that the body needs, and it is the process of metabolism that I now consider. (To describe the definition and technical details would take much space, and I refer you to the books by Schmidt-Nielsen for excellent introductions. For animals, it is the use of oxygen that is readily measured.) We know that animals operate in vastly different ways depending on their size; tiny shrews eat more than their body weight of rich insect food each day whereas elephants consume around 3 percent of their weight and that as vegetation. Is there a regular pattern to be found in the metabolism of very different animals?

Studies of metabolic rate have a long history (see the books by Schmidt-Nielsen and the full story as given by Whitfield). The most famous results are those given in 1932 by Max Kleiber (1893–1976) in his paper *Body Size and Metabolism*. An enormous number of follow-up studies have been conducted since then. Kleiber was born in Switzerland, and, after a colorful career (see Whitfield), he finished up in the Animal Husbandry Department at the University of California, Davis. Animal nutrition and needs was an important research topic. The essential results are shown in figure 8.7. The result is sometimes called the mouse-to-elephant curve, and it displays a truly remarkable regularity over about five orders of magnitude. The calculations reveal that

$$\text{metabolic rate } R \text{ scales as } R = aM^{0.74}$$
$$\text{or} \quad \log(R) = \log(a) + 0.74\log(M). \tag{8.11}$$

I have used an exponent of 0.74, but it is commonly referred to as the three-quarters exponent. It was Kleiber who first analyzed the data to produce a three-quarters exponent. He thus overturned the two-thirds law that had become dogma ever since Max Rubner did his famous measurements on dogs and other animals over many years before he retired in 1924.

The scaling results have been extended to cover a variety of other organisms. (The papers in the report edited by Brown and West give references to a large number of papers on this topic and on other scaling processes in biology.) A great deal more information has now been collected since Kleiber's pioneering work, but what has not changed for many is the strong belief in Kleiber's law: the exponent for metabolic scaling is three-quarters. But what has also not changed is the controversy around this area of biology. There are arguments about the validity of the calculations producing the three-quarters exponent, about the exponent's universality, about its accuracy, about its relevance to metabolic rates under different conditions (resting or exercising, for example), and so on. Kleiber published his book *The Fire of Life* in 1961, and now, over fifty years later, the debate about metabolism and scaling is still very much alive.

It is accepted that for many cases, metabolic rates do scale with animal size as measured by mass, and the scaling exponent may be taken as three-quarters. It is the value of the exponent, the three-quarters, that causes the problems. We might expect that the rate must scale with all of the active units in the body, so it should be proportional to mass and the exponent should be one. Or we may consider the heat generated and the mechanisms that animals use to keep temperatures constant, in which case concepts about cooling lead to metabolic rates depending on surface areas and thus to a two-thirds scaling exponent as in Rubner's old theory. In special circumstances, varying exponents may be arrived at, but it does seem that they lie between ⅔ and one, and in many cases, Kleiber's law does appear to hold. The data processing and exponent calculation by Kleiber led to the identification of an amazing regularity that few would expect to find in biology.

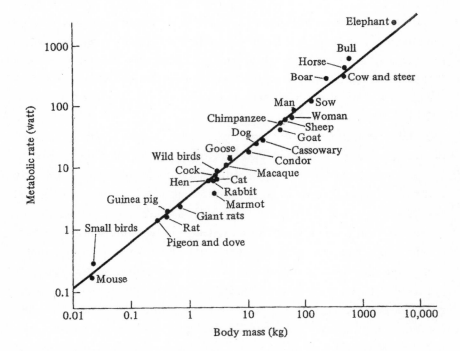

Figure 8.7. The dependence of metabolic rate on body mass for a range of animals. *Reprinted with permission, © Cambridge University Press, from K. Schmidt-Nielsen,* Why Is Animal Size So Important? *(Cambridge: Cambridge University Press, 1984).*

8.5.4 The Origin of Scaling Exponents

If the scaling process involved is simply geometric (so quantities scale, like lengths, areas, and volumes) the exponents will be ⅓, ⅔, and 1 as shown in equation (8.8). Many people accept that the calculations for metabolic rates produce exponents with confidence intervals excluding those values but centering on ¾. It also appears that exponents involving quarters rather than thirds are found in many other scaling results for different areas of biology. We now come then to the fourth role of scaling theories and must ask: Why do those particular exponents occur in nature?

The calculation of scaling exponents involving quarters has opened up a whole new research area in biology, and this might be the perfect example of George Bernard Shaw's complaint: "Science is always wrong. It never solves a problem without creating ten more."[13] A complaint for Shaw maybe, but joy for the curious scientist.

A new approach was begun by Geoffrey West and James Brown (see their 2004 *Life's Universal Scaling Laws*). They suggested that the answer lies in the networks (such as those carrying blood) that are found in biological systems and in the ways these networks are constructed and optimized. West and Brown introduced three constraints:

1. Networks service all local biologically active regions in both mature and growing biological systems. Such networks are called space filling. [Ideas from fractal mathematics are used here.]
2. The network's terminal units [capillaries, for example] are invariant within a class or taxon.
3. Organisms evolve toward an optimal state in which the energy required for resource distribution is minimized.[14]

Calculations by West and Brown (and other colleagues) have shown that concentrating on networks servicing biological systems leads to scaling exponents that involve quarters, and specifically they calculate an exponent of three-quarters for metabolic rate scaling.

The calculations, connected with biological scaling laws, have led to the discovery of biological regularities and have stimulated further research and calculations aimed at understanding the origin of these regularities. The range over which metabolic rates may be scaled is quite breathtaking, and it has been extended since Max Kleiber did his original work. I add **calculation 30, scaling from mice to elephants** as a worthy member of my list of important calculations.

(This is a very active area of research in biology, and one involving many debates and controversies. To get a flavor of the field, I have given references to some recent papers by Banavar, Glazier, Savage, and Spence as well as references for Brown and West.)

8.6 WHAT HAS BEEN LEFT OUT

As I stated at the start of this chapter, I have included only a very small sample of calculations made by scientists exploring the natural world. However, I hope this sampling has indicated the range of subjects involved and the key role that calculations can play. Maybe there should have been more about DNA and its discovery. Biophysicists will wish that I had found space for the Nobel Prize–winning work of Alan Lloyd Hodgkin and Andrew Huxley on signal transmission in nervous systems, and mathematical biologists might feel that Alan Turing's mathematically elegant work on pattern formation merited inclusion. I agree, but unfortunately there had to be some cut-off point.

Chapter 9

LIGHT

*in which we examine six calculations trying to unravel
the properties and mysterious nature of light.*

" **A** nd God said: Let there be light. And there was light." Whichever religious, mythological, or scientific version of the origins of the universe appeals to you, one thing is indisputable: one of the strangest components of our world is light. Dr. Samuel Johnson, compiler of the first dictionary, told his biographer: "We all know *what* light is, but it is not easy to tell *what* it is."[1] Johnson confronts this problem in his 1755 *A Dictionary of the English Language*, writing:

> light: that quality of action of the medium by which we see.

> Light is propagated from luminous bodies in time, and spends about seven or eight minutes of an hour in passing from the Sun to the Earth.[2]

Johnson seems to identify two aspects of light: its nature and its properties. The former is difficult to explain, and Johnson resorts to quoting some of the latter in his dictionary definition.

It is still difficult to give a definitive answer to the question about the nature of light, but the choice of theoretical concept will decide the methods used to explain its properties. In this chapter, we will see calculations relevant to both aspects, and we will conclude with a statement by Einstein that would surely have met with Johnson's approval.

9.1 THE SPEED OF LIGHT

It is clear from echoes and the rumble of distant thunder that sounds travel with a finite speed. But what of light? For many centuries, the general opinion was that the speed of light was infinite, though there were, of course, other opinions. In ancient Greece, Empedocles suggested a finite speed, but mostly authorities like Aristotle and Heron dismissed this opinion. Toward more modern times, both Kepler and Descartes believed that the speed of light was infinite, although the respected Arab physicist Alhazen (965–1040) was in the finite-speed camp, as was Christiaan Huygens. Galileo suggested that an experiment was needed; a first observer would uncover his lantern to send light to a distant second observer who, on seeing it, would uncover his lantern and send a return light to the first observer. The delay in receiving the returned signal would give the speed of light. Galileo tried it with a one-mile separation and (obviously) failed to make a satisfactory measurement. Clearly, a new approach was required.

9.1.1 Using the Moons of Jupiter

A method of determining longitude was eagerly sought in the sixteenth and seventeenth centuries. Galileo had observed moons around Jupiter, and many measurements were made of their orbits and periods in the hope that variations could be used for this illusive method. Ole Roemer (1644–1710) realized that Jupiter's moons—more precisely, its innermost moon—had another use and that was to measure the speed of light. This is another excellent example of how observations may be used in a simple calculation to give a result of major significance.

Ole Roemer was born in Aarhus on the coast of Denmark. (His name is variously spelled as Romer, Römer, and Rømer.) Although little known today, he was one of the most talented and productive men of his time. He was a scientist and mathematician of note, and he was involved in many important astronomical researches in France. If there was any justice in the world, we would celebrate Roemer rather than Fahrenheit as a pioneer designer for thermometers. He was an engineer (including in maritime matters) and studied cardioids in his quest for the optimum shape for gears. He also held civic positions in Copenhagen, such as inspector of naval architecture, purveyor of pyrotechnics and ballistics, first magistrate, mayor, and chief tax assessor. (The article by Cohen is a fascinating account of Roemer and his work. Also see Daukantas.)

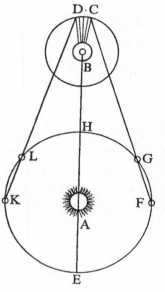

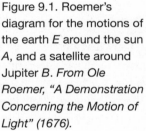

Figure 9.1. Roemer's diagram for the motions of the earth *E* around the sun *A*, and a satellite around Jupiter *B*. *From Ole Roemer, "A Demonstration Concerning the Motion of Light" (1676).*

Roemer's work on the speed of light can be readily appreciated using his original 1676 paper "A Demonstration Concerning the Motion of Light." Figure 9.1 is his diagram for explaining his method. Roemer writes:

Let *A* be the Sun, *B* Jupiter, *C* the first satellite of Jupiter, which enters into the shadow of Jupiter, to come out of it at *D*; and let *EFGHLK* be the Earth placed at divers distances from Jupiter.

Now, suppose the Earth being at *L* towards the second Quadrature of Jupiter, hath seen the first satellite at the time of its emersion or issuing out of the shadow in *D*; and that about 42½ hours after, (vid. after one revolution of this satellite) the Earth being at *K*, do see it returned in *D*; it is manifest that if the Light require time to transverse the interval *LK*, the satellite will be returned later in *D* than it would have been if the Earth had remained at L, so that the revolution of this satellite being thus observed by the Emersions, will be retarded by so much time, as the Light shall have taken in passing from L to K, and that on the contrary, in the other Quadrature FG, where the Earth by approaching goes to meet the Light, the revolutions of the Immersions will appear to be shortened by so much, as those of the Emersions had appeared to be lengthened.[3]

Roemer is saying that the innermost moon of Jupiter has a period that appears to be different depending on whether the earth is approaching or moving away from the planet, and this difference is linked to the finite speed of light.

Roemer notes that the effect is not great if observed over one moon orbit, but

that what was not sensible in two revolutions, became very considerable in many being taken together, and that, for example, forty revolutions observed on the side F, might be sensibly shorter, than forty others observed in any place of the Zodiack where Jupiter may be met with; this is in proportion to 22 minutes for the whole interval of HE, which is double of the interval that is from hence to the Sun.

Roemer has found that the speed of light is such that it takes 11 minutes to travel from the sun to the earth. He confirms that recorded observations at the Paris Observatory fit his theory. He also makes a calculation and reports on an observation made confirming his prediction:

It hath been lately confirmed by the Emersion of the first satellite observed at Paris the 9th of November last at 5 a Clock, 35' 45" at Night, 10 minutes later than it was to be expected, by deducing it from those that had been observed in the month of August, when the Earth was much nearer to Jupiter: Which M. Roemer had predicted to the said Academy from the beginning of September.

Roemer goes on to say that none of the properties of the satellite's orbit could explain the observed effect. Nothing impresses like a confirmed prediction, and from this time on, the finite speed of light was accepted. The conversion of times into speeds requires accurate values for the distances, which were not available in Roemer's time, and it is not claimed that Roemer accurately determined the speed of light—but Roemer did show how the speed of light can be measured and demonstrated the viability of his approach. (Boyer gives an entertaining discussion of the various values calculated and used by many observers in that early period of science.)

Roemer's work combining observation and calculation was innovative and clever and marks an important advance in science. **Calculation 31, light has a finite speed** is worthy of a place on the list of important calculations.

9.2 ORIGIN OF THE RAINBOW

There is something particularly delightful about rainbows. The poet William Wordsworth surely spoke for us all when he wrote:

> My heart leaps up when I behold
> A rainbow in the sky:
> So was it when my life began;
> So it is now I am a man;
> So be it when I shall grow old,
> Or let me die![4]

There is also something tantalizing and mysterious about rainbows, and they find a place in many myths and stories; for instance, the rainbow serpent plays a major part in Australian aboriginal creation stories. A rainbow is transient and shifting; you cannot go up to it and find a physical form. (Boyer gives a comprehensive history of the rainbow.)

Rainbows obviously are associated with rain, and Aristotle suggested that they are formed by light reflected from rain clouds. Over the centuries, it became apparent that a rainbow is somehow associated with the drops of water in rain itself. A theory was gradually developed to explain how rainbows occur, and it reached a modern form in the works of Descartes and Newton.

In a rainbow, we see circular arcs of colored bands in the sky as shown in figure 9.2, which is Newton's original drawing. The lower, primary bow is always seen, and as we scan it from lowest edge to highest (from E to F in Newton's drawing), it is made up of the familiar bands of color: violet, indigo, blue, green, yellow, orange, and red. The secondary bow appears higher in the sky; it is not as bright as the primary bow and not always easy to see. In the secondary bow, the order of colors is reversed, going from red to violet as we move up from G to H in Newton's figure. Over time, it was also realized that the position of the rainbow in the sky is given by precise rules: the angle between the sun's direction and the observed light from the rainbow is 42° for the primary bow (angle SEO in figure 9.2) and 51° for the secondary bow (angle SGO in figure 9.2).

Any theory for the rainbow must show its origins, explain why the colored bands occur in that particular order, and allow calculation of the characteristic angles of 42° and 51°. This provides an excellent example of how various elements of optical theory may be combined to give the framework for the calculations.

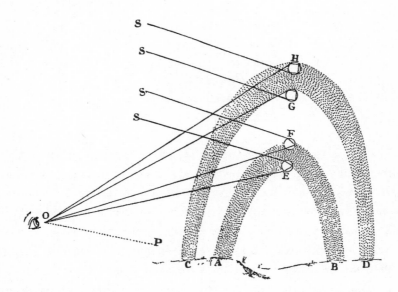

Figure 9.2. Newton's drawing of the rays of light giving rise to a rainbow. In practice, the upper, or secondary, bow is less bright than the primary bow and sometimes may not be evident at all. *S* indicates light rays from the sun, and *O* is the observer. *From Isaac Newton,* Opticks: Or, A Treatise of the Reflexions, Refractions, Inflexions and Colours of Light *(1706)*.

9.2.1 Some Basic Optics

The essential features of the rainbow may be explained by using ray optics. The principal result concerns reflection and Snell's law for refraction. In keeping with Snell's law, if a ray strikes a planar interface between two media, some light is reflected and some light is transmitted, as shown in figure 9.3. The amounts of light reflected or transmitted depend on the index of refraction n for the materials forming the interface. (For air, the refractive index may be taken as 1, and it is around 1.33 for water.) Reflection follows the law that angles of incidence and reflection are equal. Using the angles shown in figure 9.3, the angles for refraction satisfy Snell's law:

$$n_1 \sin(\theta_1) = n_2 \sin(\theta_2). \tag{9.1}$$

The second piece of information needed concerns colors. Newton used his famous prism experiment (see part 1 of book 1 of his *Opticks*) to show that white light, like that coming from the sun, comprises many different colored components. Furthermore, the index of refraction for materials like glass and water varies a little as the color of the light involved is changed. (This is why the glass prism separates out the colors.) Thus the light coming from the sun S in figure 9.2 is somehow split into its colored components to form the rainbow seen by the observer O.

9.2.2 Rainbow Formation

We assume a spherical raindrop and the path of rainbow-forming light in it is shown in figure 9.4. Rays from the sun refract into the raindrop and reflect at the inner surface according to the rules introduced above (the sphere is assumed to be locally plane at the points of contact so those rules may be applied). The rays then leave the raindrop by another refraction event. Figure 9.4 (a) shows how light forming the primary bow reflects once inside the raindrop, while (b) shows the mechanisms for the secondary bow involving two reflections. Since there is some transmission also at a reflection point, the light is diminished in intensity at each reflection, and thus we find that with two reflections involved, the secondary bow is less intense than the primary. From figure 9.4 (a), we see that it is the angle γ that measures the observed angle between the sun's rays and those seen by the observer.

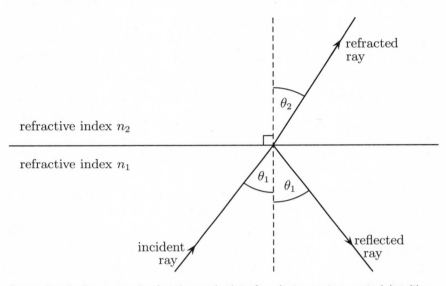

Figure 9.3. Reflection and refraction at the interface between two materials with refractive indices n_1 and n_2. The incidence angle is θ_1, and the refracted angle is θ_2. *Figure created by Annabelle Boag.*

The ray incident on the raindrop forms the angle α; then the refraction angle β, and the rainbow angle γ follow from the laws of optics and a little geometry as

$$\sin(\alpha) = n \sin(\beta) \; [n = \text{refractive index of water}] \qquad \gamma = 2(2\beta - \alpha). \quad (9.2)$$

Finding the appropriate angle α is a technical problem that involves first considering all possible rays and then identifying those that give the dominant contribution to the rainbow. This leads to the caustic ray (details may be readily found in the fine expository article by Casselman and in the paper by Nussenzveig). The result is that we must set

$$\sin(\alpha) = \sqrt{\frac{4-n^2}{3}} \quad \text{which for } n = 1.33 \text{ gives} \quad \alpha = 59.6°.$$

Using equation (9.2) leads to the internal angle $\beta = 40.4°$ and then to a rainbow angle γ of $42.4°$.

Thus it was that Descartes and Newton explained the mechanism behind the rainbow and calculated the observed angle of $42°$ for the primary bow. Using similar ideas and figure 9.4 (b) allows the rainbow angle for the secondary bow to be calculated as $51°$ matching that observed.

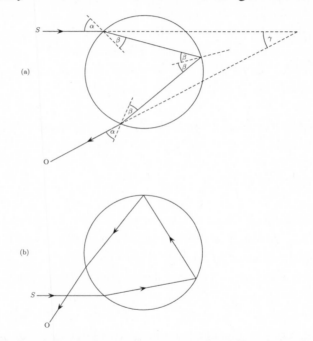

Figure 9.4. Paths of rays as they encounter a raindrop. (For clarity, only the relevant refracted or reflected rays are shown at the boundaries, and the other reflections or transmissions that can occur there are not shown.) The incident angle is a, and the reflection angle inside the drop is β. The rays in (a) go from the incident sunlight S to the observer O to form the primary bow, and (b) shows how the secondary bow is formed through two reflections. In (a), γ is the rainbow angle—the angle between the direction of the sun's rays and those going from the rainbow to the observer. *Figure created by Annabelle Boag.*

9.2.3 Rainbow Colors

You may say that is all very well and an impressive calculation of the rainbow angles, but where do the colors come in? Newton had discovered that the refractive index of materials varies according to the color of the light involved. For water, the refractive index decreases as the wavelength of light increases; $n = 1.34055$ for blue light of wavelength 450 nm, and $n = 1.33257$ for red light of wavelength 650 nm. Carrying out the calculations translates this information into a spread of rainbow angles γ, and so each color produces its own band in the rainbow. Furthermore, looking at figure 9.4 (b), it is seen that a more extensive path is followed by the rays giving the secondary bow, and this time the direction of the spread of rainbow angles is reversed, so matching the observed phenomena.

These results can be given in precise form for the size of the rainbows. This is how Newton himself reported it in his *Opticks*:

> The breadth of the interior bow *EOF* [see figure 9.2] measured across the colours shall be 1 Degr. 45 Min. and the breadth of the exterior *GOH* [the secondary bow as in figure 9.2] shall be 3 Degr. 10 Min. and the distance between them *GOF* shall be 8 Degr. 15 Min.[5]

These calculations represent a triumph for the theory of optics, and **calculation 32, seeing a rainbow**, certainly deserves a place in my list of important calculations. The initial theory of the rainbow has been built on to explain other features such as the faint green and pink bands, known as supernumerary arcs, which may sometimes be seen beneath the primary bow. (The article by Nussenzveig is good place to start for further details.) There are also similar phenomena, like glories and halos, that may be analyzed using rainbow optics.

9.2.4 A Different Response

A scientist reading about the work of Descartes and Newton might say "what a beautiful result," but others see it in a quite different way. I began this chapter by mentioning the allure of the rainbow and its magical beauty. For some people, the seeming reduction of the rainbow to angles in ray optics and variations in the refractive index of water destroys the magic, the sense of awe, and the beauty. This was most famously expressed by the poet John Keats (1795–1821) in his poem "Lamia":

> Do not all charms fly
> At the mere touch of cold philosophy?
> There was an awful rainbow once in heaven:
> We know her woof, her texture; she is given
> In the dull catalogue of common things.
> Philosophy will clip an Angel's wings,
> Conquer all mysteries by rule and line,
> Empty the haunted air, and gnomed mine—
> Unweave a rainbow.[6]

The interaction of science and poetry is complex (interested readers might consult the book by Marjorie Hope Nicolson and the book edited by Heath-Stubbs and Salman). For example, John Donne (1571–1631) wrote:

> And the new philosophy puts all in doubt,
> The element of fire is quite put out;[7]

James Thomson (1700–1748) wrote a flattering poem entitled "To the Memory of Sir Isaac Newton," which includes the lines:

> Even now the setting sun and shifting clouds,
> Seen, Greenwich, from thy lovely heights, declare
> How just, how beauteous the refractive law.[8]

Personally, Wordsworth's words given at the start of this section, still resonate with me, but, equally, I feel a certain thrill when I see Newton's brilliant calculations.

9.2.5 Light Rays

You may have noticed that when I introduced the ray theory of light in section 9.2.1 I was talking about how light behaves and that was enough to give us the theory of the rainbow. I did not say anything about the nature of a ray of light. Here is how Isaac Newton begins his *Opticks*:

> Definition I. By the Rays of Light I understand its least Parts, and those as well Successive in the same Lines, as Contemporary in several Lines. . . . The least Light or part of Light, which may be stopped alone without the rest of the Light, or propagated alone, which the rest of the Light doth not or suffers not, I call a Ray of Light.[9]

I will have more to say on this later.

9.3 WAVE THEORY AND LIMITS ON VISION

The theory of the rainbow provides a good example of how Newton could use his rays of light and spectrum of colors to explain a great variety of optical phenomena. His *Opticks* is a masterpiece of physics. However, for some optical effects, his explanations seem contorted, and some strange properties must be introduced for his particles, or corpuscles, of light and rays. At times, Newton is almost introducing elements of the main competitor for his theory of optics: wave theory as set out by Christiaan Huygens. Newton's contemporary, Christiaan Huygens (1629–1695) was a brilliant scientist; unfortunately, and perhaps unfairly, he is overshadowed by Newton. Huygens published his *Treatise on Light* in 1690, setting out a wave theory for light and explaining how light is propagated, reflected, and refracted. This work might be viewed either as a rival to Newton's ray theory or as complementary to it.

Huygens wrote, "I have shown in what manner one may conceive light to spread successively, by spherical waves, and how it is possible that this spreading is accomplished with as great a velocity as that which experiments and celestial observations demand."[10] His central idea is shown in figure 9.5. It shows how various parts of a candle flame emit spherical light waves to form the total candlelight. The second drawing shows how a point source at *A* emits light in a spherical wave and how Huygens describes light propagation. If there is a wavefront at some place in the spread of light away from a source, each point on that wavefront may be thought of as the source of a new spherical wave; and the waves from all those points combine to give the new wavefront of the propagating light wave. Thus the whole wavefront *DCEF* is generated by all the points *d* on the preceding wavefront *L*. This is known as Huygens's principle, and it is a central part of the theory of optics. If the wavefront is planar, the secondary sources on it give individual spherical waves that combine to form the next propagating plane wavefront. Thus using this principle, Huygens could describe the propagation of light in a great variety of circumstances.

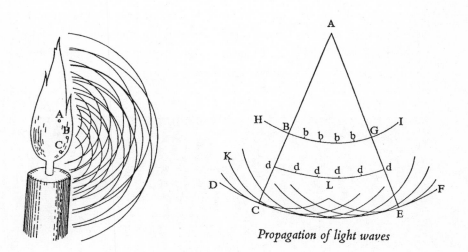

Propagation of light waves

Figure 9.5. Spreading of light waves as described by Huygens.
From Christiaan Huygens, Treatise on Light *(1690).*

Using these ideas, Huygens showed how wave theory explains various phenomena in optical reflection and refraction. This included his chapter 5, "On the Strange Refractions of Iceland Crystal," an optical effect we now know as double refraction.

Following Newton and Huygens, there was a vigorous investigation of optical phenomena and much debate about the appropriate theory for light (the book by Cantor is recommended). One of the leaders in these endeavors was Thomas Young (1773–1829). (Young was a polymath, and the aptly named book by Robinson gives a very readable account of his life.) In particular, Young investigated the interference and diffraction of light, two optical effects that were difficult to reconcile with Newton's particle theory of light. Young's famous double-slit interference experiment is a classic in optical science (see, for example, chapter 4 in Heavens and Ditchburn).

The proponents of the wave theory of light had the advantage as they could use ideas coming from the theory for sound and water waves. With these theories, it was very clear what was happening, and it was easy to see how waves could spread around obstacles and diffract from apertures. (There was a problem, of course: the media carrying those sound or water waves were well known, whereas if an underlying medium was sought for optical waves, it led back to unresolved questions about an ether, a medium that had to have very strange properties.)

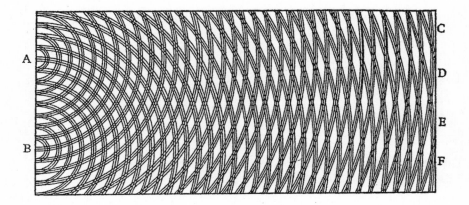

Figure 9.6. Interference of water waves as depicted by Thomas Young. The sources are *A* and *B*, and Young pointed out that the interference at *C*, *D*, *E*, and *F* meant that the water was nearly smooth there. *From Wikimedia Commons, user Sakurambo.*

Young did not do extensive mathematical calculations, but he explained things using what we might call an analogue computer. Around 1802, Young set up a ripple tank which allowed water waves to be created and observed in a variety of configurations (see Robinson chapter 7). In one particularly valuable case, he demonstrated how waves from two sources could create an interference pattern, as shown in figure 9.6.

Gradually a whole body of experimental work was carried out to support the ideas propounded by Huygens, Young, and others supporting the wave theory. Huygens's principle could be used to understand many of these effects, and Young's analogue demonstrations proved to be very convincing. What was lacking was a solid mathematical foundation for the wave theory of light.

The mathematical theory was developed in the nineteenth century with Augustin-Jean Fresnel (1788–1827) playing a leading part. Essentially Fresnel gave a mathematical form to Huygens's principle. Suppose we have a light field E_0 given in the xy-plane as shown in figure 9.7. Then an element in that plane contributes to the light E at observation point P an amount

$$dE = \text{constant} \times E_0 \frac{\exp(-2\pi iR/\lambda)}{R} \, dxdy. \qquad (9.3)$$

This is just the mathematical form of Huygens's spherical wave contribution from an element of the initial wave. To find the total light wave at P will require integration over all the elements making up the initial light wave. An aperture has been indicated in figure 9.7, since that is the usual case in practice, and then the integration is carried out over that aperture.

Carrying out the integration is not so simple unless approximations are made appropriate to the physical or experimental arrangement being used. The cases known as Fraunhofer diffraction and Fresnel diffraction are commonly considered. (See Heavens and Ditchburn chapter 6, or Lipson and Lipson chapter 7.) The result is a beautiful set of patterns illustrating the consequences of light behaving like a wave.

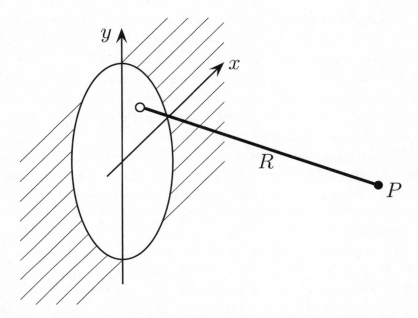

Figure 9.7. Geometry for Fresnel's optical wave theory. An element in the source plane contributes to the light wave at point *P*, which is a distance *R* from that element. *Figure created by Annabelle Boag.*

9.3.1 The Airy Disc

The matching of theory and experiment for diffraction and interference effects was central to the proof that light behaves like a wave. However, from a practical point of view, it is the consequences of the wave-like nature of light for the operation of imaging systems (such as our eye, a telescope, or a microscope) that assumes a vital importance. All practical systems (like those shown in figure 9.8) will involve an aperture, and this will limit the extent of the incident wave that the lens delivers to the observation point *P* in the image. The effects of the aperture are exactly the kind of diffraction effects first examined by those nineteenth-century optical physicists mentioned above.

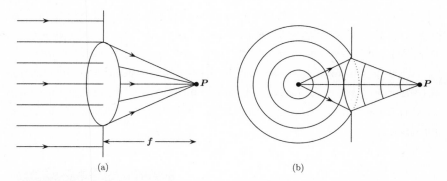

Figure 9.8. Effects of apertures in imaging systems. (a) The case for long-distance observation with a telescope giving an image in the focal plane. (b) How some waves from an object are lost as it is imaged at *P*. *Figure created by Annabelle Boag.*

Viewing stars through telescopes led the Astronomer Royal George Biddell Airy (1801–1892) to calculate the effect of the telescope aperture on the nature of the observed image. In his 1834 paper "On the Diffraction of an Object-Glass with Circular Aperture," he writes that

> The image of a star will not be a point but a bright circle [he should really say disc] surrounded by a series of bright rings. The angular diameters of these will depend on nothing but the aperture of the telescope, and will be inversely as the aperture.[11]

Airy goes on to calculate what is now known as the Airy diffraction pattern and to discuss how it limits the details that may be seen using a telescope. In modern terms, the intensity of light I at a point a distance r from the center of the pattern is given by

$$I(r) = I_0 \left[\frac{J_1(\rho)}{\rho} \right]^2 = I_0 \left[\frac{J_1(2\pi Rr / \lambda f)}{2\pi Rr / \lambda f} \right]^2 . \qquad (9.4)$$

In equation (9.4) I_0 is a constant, R is the aperture radius, f is the lens focal length, and λ is the wavelength of light. (J_1 is the Bessel function of order one. There will be more on Bessel functions in chapter 12.) The result is plotted in figure 9.9 where an example of the appearance of an Airy pattern is also shown (with the outer rings somewhat enhanced).

This might have remained an astronomical specialty, but in fact this diffraction effect is the limiting factor in all imaging systems, so it is also vitally important for microscopes and for imaging systems such as the human eye.

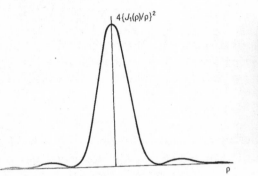

Figure 9.9. A plot of the intensity of light in the Airy diffraction pattern and an experimental example. *Reprinted with permission of John Wiley & Sons, from O. S. Heavens and R. W. Ditchburn,* Insight into Optics *(Hoboken, NJ: John Wiley & Sons, 1991).*

9.3.2 Resolution of Image Details

If images are blurred by diffraction effects, we need to ask what level of detail in an object can still be resolved in its image. Using Rayleigh's criterion (named after its originator Lord Rayleigh) is a widely accepted way to proceed. Rayleigh suggested that two points in an image may be said to be just resolved if the Airy disc center of one of them falls on the first zero radius for the second one, as shown in figure 9.10. Images of points closer than that look like one blurred big point, whereas at that separation or greater, two distinct image peaks may be discerned. In terms of a resolved angle β_r, equation (9.4) leads to the famous formula in terms of the wavelength of light involved and the aperture diameter D:

$$\beta_r = 1.22 \frac{\lambda}{D} \quad \text{radians.} \tag{9.5}$$

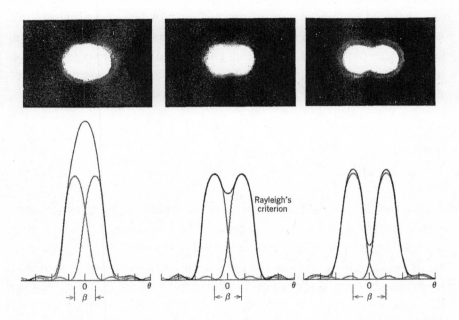

Figure 9.10. Diffraction patterns (Airy discs) for two point sources imaged with varying closeness. *Reprinted with permission of John Wiley & Sons, from D. Halliday and R. Resnick,* Fundamentals of Physics, *2nd ed. (Hoboken, NJ: John Wiley & Sons, 1991).*

For the human eye in bright light (so D is about 2 mm) and light of wavelength 500 nm, the angle of resolution is around one minute of arc. The optics of the eye is quite complicated, but the pupil does provide an aperture in the visual system. (The book by Wandell gives experimental data for the form of the blurred image on the retina; see chapter 2, figure 2.5.) Because there is a resolution limit forced by the imaging system, it would be pointless to have an arrangement of the photoreceptor cells in the retina of the eye (known as the photoreceptor mosaic) to sample the image beyond the level of detail present as shown in figure 9.10. Thus the matching of optics and image-sampling photoreceptor density can be observed as discussed by Denny and McFadzean (see bibliography).

In section 6.4.3, we saw that George Biddell Airy was often portrayed as the villain in the story of planet Neptune's discovery. However, most people would know him for the Airy disc and the ideas about limits to the details that may be extracted from an image. He has redeemed himself then with his part in the highly significant **calculation 33, diffraction and the limit to vision**. It is the fact that light behaves like waves, and there is diffraction from apertures, that limits our visual capacity.

9.4 FROM OPTICAL WAVES TO ELECTROMAGNETIC WAVES

The wave theory of light was enormously successful for describing a large range of optical phenomena. However, there remained the underlying mystery: Physically, what are these waves? It was clear that the surface of water moves up and down as a water wave travels along that surface, and sound waves are related to pressure variations in the air. For many people, the existence of a strange medium—the ether—was required to give some sort of mechanical picture for light waves. The story of the ether and its eventual banishment is an intriguing part of the history of physics. A major step forward was made by James Clerk Maxwell in the second half of the nineteenth century.

Maxwell (1831–1879) was the greatest theoretical physicist of the nineteenth century, making major contributions to many areas of physics. (There are several biographies of Maxwell—I particularly like those by Harmon and Mahon, and Segrè gives a good introduction to his work.) Maxwell was born in Edinburgh and educated at Cambridge University. He would go on to become a professor of physics, first at Aberdeen and then at Kings College, London. Maxwell had a lifelong interest in electricity and magnetism and the part that a hypothetical ether might play in it. His magnificent 1873 *Treatise* was the first comprehensive treatment of the subject.

9.4.1 The Phenomena of Electricity and Magnetism

The work of André-Marie Ampère, Charles-Augustin de Coulomb, Michael Faraday, Hans Christian Oersted, and others gradually built up knowledge of electrical forces, magnetic phenomena, and interactions between electric and magnetic elements. The force between charges acts over a distance and was shown to follow an inverse square law. Faraday showed that moving a magnet in and out of a loop of wire generated an electric current in that wire; this is magnetic induction and the origin of the electric generator. Conversely, a current flowing in a wire moves a magnetic needle brought close to that wire. Faraday's experiments with iron filings showed that a magnet can influence objects around it but not in contact with it. It was clear that electric and magnetic effects could occur without the need for electric charges or magnets to be in contact. Thus the influence of charges and magnets act over a distance, and constants were introduced to characterize the nature of the intervening medium, which might be air, water, or glass, for example. These constants are the dielectric constant (or permittivity) ε for electrical effects and the magnetic permeability μ for magnetic effects. In free space, these are denoted by ε_0 and μ_0. For example, the force between charges q_1 and q_2 separated by a distance r has magnitude $(q_1 q_2)/\varepsilon r^2$.

Maxwell introduced the electric field E to describe the influence of charges in the space around them and the magnetic field H for similar magnetic effects. These fields are vector quantities and depend on the position in space and change with time as the charges and magnets move, so we should write $E(x, y, z, t)$ and $H(x, y, z, t)$. Taken together, we refer to the electromagnetic field. Maxwell then demonstrated his genius by writing down a set of equations showing how electric and magnetic fields are linked and how they are generated by charges and electric currents. These are the theoretical equivalent of the experimental picture built up by Faraday. It is the form of these equations in a region of space devoid of charges and currents that concerns us here. In a uniform medium (with the dielectric constant the same everywhere) the electromagnetic fields must satisfy the following equations:

$$div\ \mathbf{E} = \nabla.\mathbf{E} = 0 \qquad\qquad div\ \mathbf{H} = \nabla.\mathbf{H} = 0$$

$$curl\ \mathbf{E} = \nabla\times\mathbf{E} = -\mu\frac{\partial\mathbf{H}}{\partial t} \qquad curl\ \mathbf{H} = \nabla\times\mathbf{H} = \varepsilon\frac{\partial\mathbf{E}}{\partial t}. \quad (9.3)$$

These are known as Maxwell's equations. It is not necessary to understand the mathematics in detail, but you should appreciate that these wonderful equations do tell us how the electromagnetic fields can vary in space and time. You can also see how E and H are mixed together in the same equation to give Faraday's magnetic induction. (The books by Heavens and Ditchburn, Lipson and Lipson, and Segrè give further details with particular reference to optics.)

9.4.2 Maxwell's Miraculous Calculation

Maxwell manipulated his equations (see the aforementioned references) to show that the electromagnetic fields in a charge-free and magnet-free uniform isotropic medium must satisfy

$$\nabla^2\mathbf{E} = \varepsilon\mu\frac{\partial^2\mathbf{E}}{\partial t^2} \qquad \text{and} \qquad \nabla^2\mathbf{H} = \varepsilon\mu\frac{\partial^2\mathbf{H}}{\partial t^2}. \quad (9.4)$$

Again, you need not appreciate the mathematical details of these equations, but you should recognize their form: these are wave equations. Maxwell had shown that the electromagnetic field could take the form of waves in space; electromagnetic disturbances, or fields, can propagate in space in the form of waves.

Faraday had speculated about the links between light and electromagnetism. For example, in what is now called the Faraday effect, he showed that a magnetic field could change the polarization of a light wave. He also wondered about how radiation and vibrations of lines of force might be connected. Faraday clearly had the ideas about a link between optics and electromagnetism, but he lacked the mathematical background to establish the full picture. It was Maxwell who was ready and able to take the definitive step; it involved showing that his electromagnetic waves are the same as light waves. According to the standard wave equation, the waves described by equations (9.4) have speed V given by

$$\text{wave speed,} \quad V = \frac{1}{\sqrt{\varepsilon\mu}} \quad \text{and in free space the speed is} \quad V = c = \frac{1}{\sqrt{\varepsilon_0\mu_0}}. \quad (9.5)$$

However, Maxwell knew that there was another quantity with the dimensions of a velocity but which was related to definitions of electrostatic and electromagnetic units. He called this quantity v. Most importantly, v could be found experimentally using electric circuits (see Segrè appendix 6, or Lipson and Lipson section 4.2.2, for a simple introduction). The electromagnetic theory for those circuits gave $v = 1/\sqrt{\varepsilon\mu}$, exactly the same form as he had derived for the wave speed V. I will let Maxwell tell the amazing story (as it is written in his *Treatise*), and you should note that he is referring to the case of free space or waves and the electronic circuits in air.

On the theory that light is an electromagnetic disturbance, propagated in the same medium through which other electromagnetic actions are transmitted, V must be the velocity of light, a quantity the value of which has been estimated by several methods. On the other hand, v is the number of electrostatic units of electricity in one electromagnetic unit, and the methods of determining this quantity have been described in the last chapter. [See the references given above.] They are quite independent methods of finding the velocity of light. Hence the agreement or disagreement of the values of V and of v furnishes a test of the electromagnetic theory of light.

In the following table the principal results of direct observation of the velocity of light, either through the air or through the planetary spaces, are compared with the principal results of the comparison of the electric units:

Velocity of Light (mètres per second).		Ratio of Electric Units. (metres per second)	
Fizeau	314000000	Weber	310740000
Aberration, etc and Sun's Parallax	308000000	Maxwell	288000000
Foucault	298360000	Thomson	282000000

It is manifest that the velocity of light and the ratio of the units are quantities of the same order of magnitude.[12]

(In the version of Maxwell's *Treatise* cited in the bibliography, there is an extended table of V and v values that were available in 1889; the equality of V and v is apparent.)

Of course, Maxwell has now been fully vindicated, and it is generally accepted that light waves are indeed electromagnetic waves, and Maxwell's equations are the appropriate vehicle for theoretically investigating their behavior. Maxwell's calculation and comparison of those velocities obtained from such wildly different methods ranks as one of the greatest achievements in science and no list of important calculations would be complete without **calculation 34, light and electromagnetism**.

9.4.3 Triumphs and New Mysteries

If light is an electromagnetic wave, we have an immediate explanation of things like the Faraday Effect (a magnet influencing a light wave). Also, as Maxwell pointed out in his *Treatise*, the form of the velocity V in equation (9.5) tells us how light behaves in different materials (through their values of ε and μ) and leads to the introduction of the refractive index. As mentioned in the previous section, light has a polarization, meaning that its waves are transverse. Using Maxwell's equations (9.3), it is easy to show that electromagnetic waves must have their E and H perpendicular to the direction of propagation; electromagnetic waves are transverse waves and thus explain the polarization of light.

We should also note that Maxwell's theory covers all electromagnetic waves—not just those we know as light, but also things like x-rays and radio waves.

Maxwell's electromagnetic theory of light together with Newton's laws of motion form the basis for classical physics, which reached its triumphal peak at the end of the nineteenth century. But what about that old mystery: What exactly is light? The answer—it is an electromagnetic wave—only brings out the obvious question: Physically, what is an electromagnetic wave? And there we remain stuck. (A wonderful discussion by the matchless expositor Richard Feynman can be found in section 20-3, "Scientific Imagination," of *The Feynman Lectures on Physics*.) One approach to the conceptual difficulty encountered when discussing electromagnetic waves is to just use Maxwell's theory to calculate the results of observations and experiments. In effect, we simply follow the opinion of Heinrich Hertz (the man who first demonstrated the existence of radio waves): "Maxwell's theory is Maxwell's system of equations."[13]

9.5 BACK TO PARTICLES?

We have just seen that the combination of Newton's laws of motion and Maxwell's equations for electromagnetism covers what today we call classical physics. Some people thought that they covered virtually all of physics. Here is the famous statement made early in the twentieth century by the eminent experimental physicist Albert Abraham Michelson (1852–1931):

> The more important laws and facts of physical science have all been discovered, and these are now so firmly established that the possibility of their ever being supplanted in consequence of new discoveries is exceedingly remote. Nevertheless, it has been found that there are apparent exceptions to most of these laws, and this is particularly true when the observations are pushed to a limit. . . . Many other instances might be cited, but these will suffice to justify the statement that "our future discoveries must be looked for in the sixth place of decimals."[14]

It is ironic that Michelson's experiments designed to detect the ether played a part in Einstein's thinking about the theory of relativity, which, together with the introduction of quantum theory, revolutionized physics in the twentieth century.

Serious problems began to arise when physicists tried to account for the spread over wavelengths of radiation coming from very hot bodies. In trying to derive the observed radiation law using Maxwell's theory and classical statistics (which had been successful in deriving the properties of gases—see the next chapter), scientists like Sir James Jeans and Lord Rayleigh ran into unexpected difficulties. In some cases, they were getting infinite answers for the (obviously finite) amount of radiation emitted—hardly a problem "in the sixth place of decimals"! People spoke of the ultraviolet catastrophe. (For a very readable introduction to this, see Gamow's lovely book. The book by Pais is also a wonderful account of these events and Einstein's contributions. See Rigden for a briefer introduction.)

The leading figure in the field of radiation was Max Planck who looked at radiating "black bodies" in terms of discrete oscillators rather than continuous waves. In this way, he was able to bring order back to the field. The whole area of radiation theory and optics was completely changed by Albert Einstein's 1905 paper "On a Heuristic Point of View about the Creation and Conversion of Light." Einstein wrote that

> according to the assumption to be contemplated here, when a light ray is spreading from a point, the energy is not distributed continuously over ever-increasing spaces, but consists of a finite number of energy quanta that are localized in points in space, move without dividing, and can be absorbed or generated only as a whole.[15]

Later, these "energy quanta" came to be called photons, and the quantum theory of light was born. Light of wavelength λ has frequency v given by frequency = speed of light divided by wavelength $v = c/\lambda$. For this light, each photon carries the same energy and momentum:

$$\text{energy } E = hv \quad \text{and momentum } p = \frac{hv}{c}. \tag{9.6}$$

Equation (9.6) contains h, the newly introduced quantum constant, now known as Planck's constant. The modern value for h is 1.0546×10^{-27} erg-sec, and its remarkably small size tells us that we have entered the quantum world of very small things.

Einstein quite rightly used the term "very revolutionary"[16] to describe his 1905 paper launching the quantum theory of light. But the very small value of h means that everyday light involves an enormous number of photons, and, generally, we are quite unaware of this discrete nature of light. The question inevitably arises: How do we know that photons actually exist? The answer requires us to once again appreciate that vital link between theory and experiment, and it is the calculations involved that form my choice of **calculation 35, photons exist**.

9.5.1 Einstein Knocks Out Electrons with Photons

In his 1905 paper, Einstein considered three topics which could profitably be discussed using his new quantum theory of light. The most convincing for many people was Einstein's theory of the photoelectric effect. (The other two topics were Stokes's rule in photoluminescence and the ionization of gases by ultraviolet light.) The photoelectric effect was discovered in 1887 by Heinrich Hertz (of radio-waves fame) and subsequently investigated by Philipp Lenard in 1903 and many other experimentalists. It refers to the emission of electrons from a metal surface when light is incident on it. There are three key observations for the photoelectric effect:

1. The energy of the individual photoelectrons is independent of the intensity of the light.
2. The number of emitted electrons (the photoelectric current) is proportional to the intensity of the light.
3. For a given metal, electrons are emitted only when some threshold frequency of light is exceeded, and then their energy increases linearly as frequency is increased.

These results are shown graphically in figure 9.11.

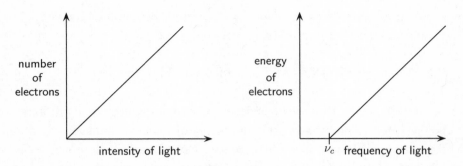

Figure 9.11. Results for the photoelectric effect. Number of photoelectrons emitted from a metal surface versus light intensity, and emitted electron maximum energy versus light frequency. No electrons are emitted when the frequency is less than the threshold frequency v_c. *Figure created by Annabelle Boag.*

Einstein did a simple calculation to show a likely origin for these facts. He assumed that a given metal had a characteristic work P that must be done on its electrons in order to displace them from the metal surface. In the light of frequency v incident upon the metal, there will be photons with energy hv. If these photons hit the electrons, they can give them energy to overcome the required work P and have them emitted with energy E according to

$$E = hv - P. \tag{9.7}$$

With equation (9.6) Einstein shows that electrons can escape the metal with positive energy E only when the frequency is large enough to make $hv - P$ greater than zero. Beyond that cut-off point, equation (9.7) tells us that the energy of the electron E increases linearly with the frequency v.

If, for a fixed frequency v, the intensity of the light is increased, it will consist of photons each with energy hv but there will be more of them. This means that there will be more photons to cause more electrons to be emitted from the metal in accordance with equation (9.7). Thus increasing the intensity of light should not change the energy of the individual emitted electrons, but it should increase their number.

By 1916, Robert Millikan had experimentally demonstrated the accuracy of Einstein's theory (his results for electron energy versus radiation frequency are reproduced by Rigden), but he was not convinced by the underlying assumptions and wrote that

> despite then the apparently complete success of the Einstein equation, the physical theory on which it was designed to be the symbolic expression is found so untenable that Einstein himself, I believe, no longer holds to it, and we are in the position of having built a very perfect structure and then knocked out entirely the underpinning without causing the building to fall.[17]

Millikan was wrong about Einstein, and the validity of the light quanta concept was becoming widely accepted. With his little calculation, Einstein has completely explained the three key observations on the photoelectric effect. As a measure of the importance of this work, we can note that when Einstein was awarded the 1922 Nobel Prize it was given to Albert Einstein "for his services to theoretical physics and especially for his discovery of the law of the photoelectric effect."[18] (If you are surprised that the prize was not given for Einstein's work in relativity, there is an interesting background to the subject—see Pais chapter 30.)

9.5.2 Compton Changes Photons with Electrons

If the electron is not tightly bound in some way, when a photon hits it, there will be a scattering process, and the output will be another photon and an electron moving in some particular way. This process is known as Compton scattering, and the dynamics involved are shown in figure 9.12. In this particular process, the radiation used in experiments is not light but x-rays, and, for x-rays, the electrons are loosely bound in ordinary atoms. The dynamics of the process are worked out using the conservation laws for energy and momentum. The explanation of the photoelectric effect is basically an argument about energies, but now the photon momentum as in equation (9.6) must also be used. (For more details and history see the books by Gamow and Williams.)

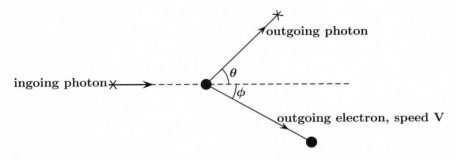

Figure 9.12. Compton scattering: the collision of a photon with a stationary electron. *Figure created by Annabelle Boag.*

If the incoming photon has frequency v_0, and the outgoing, or scattered, photon has frequency v, according to equation (9.6) the electron gains an amount of energy $hv_0 - hv$. We must also make sure that the momentum is conserved in the two directions involved, that is the direction of the incoming photon and the direction perpendicular to that, or the horizontal and vertical directions in figure 9.12. Letting the outgoing electron have mass m and speed V, and using equation (9.6) for the photon momentum, leads us to

$$\frac{hv_0}{c} = \left(\frac{hv}{c}\right)\cos(\theta) + mV\cos(\phi),$$

$$0 = \left(\frac{hv}{c}\right)\sin(\theta) - mV\sin(\phi).$$

Using these conservation-laws equations allows us to calculate the result of the scattering process. The startling conclusion is that the outgoing photon has a different frequency for different scattering angles θ (as in figure 9.12). The result is usually expressed in terms of wavelengths λ rather than frequency v with $\lambda = c/v$. The outgoing photon has wavelength λ larger than the incoming photon wavelength λ_0 by an amount

$$\lambda - \lambda_0 = \left(\frac{h}{mc}\right)\left[1 - \cos(\theta)\right]. \tag{9.8}$$

Equation (9.8) tells us that if we look at the scattered photons at various angles, we will see a wavelength change varying from zero at $\theta = 0$, to (h/mc) at $\theta = 90°$, to ($2h/mc$) at $\theta = 180°$.

The calculation of the effects of radiation-electron scattering using the photon-electron picture comes up with a clear prediction involving the quantum constant h. The detailed experiments carried out by Arthur Holly Compton in 1922 confirmed the results of the photon theory, and, in 1927, he was awarded the Nobel Prize for his work. In his 1923 *Physical Review* paper, Compton wrote that his results convinced him "that the radiation quantum carries with it directed momentum as well as energy."[19] So **calculation 35, photons exist** is based on work that merits not one but two Nobel Prizes! But does it really answer the question about whether photons give us the true nature of light? I return to that in section 9.7.

9.6 THE BENDING OF LIGHT

If light consists of particles, as Newton supposed, it is natural to ask whether these particles are affected by gravitational forces. In fact, Newton raised the matter in his *Opticks*:

> Query 1. Do not Bodies act upon Light at a distance, and by their action bend its rays; and is not this action (caeteris paribus) [other things being equal] strongest at the least distance?[20]

Finding the answer to this question has proved to be of great importance in the development of science.

9.6.1 The First Deliberations

In section 7.5.2, we saw that the astronomer Rev. John Michell followed Newton's suggestion and showed that the assumption of a gravitational pull on light led to the idea of a "dark star." In 1783, Michell wrote to Henry Cavendish (1731–1810) (famous for his measurement of the gravitational constant), suggesting gravity might reduce the speed of light, so that a method for measuring the mass of a star might be devised by exploiting that effect. That stimulated Cavendish to calculate by how much the path of light might be bent as it closely passed by a massive body. Cavendish never published his calculation. (For further details, see the book and 1988 paper by Will.) Pierre-Simon Laplace also wrote about the possibility of a "dark star," and this probably stimulated the Bavarian astronomer Johann Georg von Soldner (1776–1833) to independently calculate the bending of a light ray as it passed by the sun. Soldner did publish his findings in 1804.

The physical situation is shown in figure 9.13. The problem is to calculate the bending angle θ when the light ray passes a distance Δ from the center of the sun. Clearly, the maximum deviation occurs when the ray just grazes the sun and Δ is equal to the sun's radius.

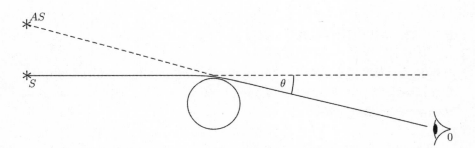

Figure 9.13. The path of the light from a star S is bent through an angle θ as it passes the sun. An observer O would interpret the light as coming from the apparent star AS. The direct line from S to O may pass through the sun so S would not be observed if it was not for the illustrated bending phenomenon. *Figure created by Annabelle Boag.*

If we treat the particles of light like any other particle, the mathematical problem is to solve their equation of motion (from Newton's second law),

$$\frac{d^2\mathbf{r}}{dt^2} = \frac{GM\mathbf{r}}{r^3},$$

(9.9)

for the position vector $\mathbf{r}$ of the particle as it moves past the mass M situated at $\mathbf{r} = \mathbf{0}$. Solving equation (9.9) is a standard problem, which gives the elliptical orbits for planets as discussed in chapter 6. For the situation in figure 9.13, the particle is moving so fast that it goes by the sun in the same way that we get fly-by data for a spacecraft moving past a planet. The orbit is a hyperbola (see the 1988 paper by Will for details) and gives the light bending as $\theta = 0.875"$. Soldner gives 0.84" in his 1804 paper. These are very small angles, and so experimental tests seemed most unlikely.

(A technical aside: Equation (9.9) involves the mass M of the sun, but not the mass of the particle passing by it. This is because the inertial mass and the gravitational mass are equal, and so the particle mass occurs on both sides of the equation of motion and thereby may be cancelled out. Thus no question about the mass of the particles of light is involved.)

9.6.2 Einstein Changes the Physics

Einstein's definitive paper *The Foundations of the General Theory of Relativity* appeared in 1916. In it he argued that Newtonian mechanics, based on the interaction of bodies by means of forces, should be replaced by a new theory in which gravitational effects manifest themselves by changing the metric properties of space around bodies. Newton's theory emerges as the first approximation to Einstein's theory, and in the very simplest case, the equation to be solved reduces to Newton's equation (9.9). It is only in extreme cases (for example, very massive bodies and speeds approaching the speed of light) that the results from Einstein's theory differ substantially from those given by Newton. (The books by Lambourne and Weinberg are recommended for the technical details.) The phenomenon of light bending as shown in figure 9.13 offers one case where those extreme effects might come into play.

Some details of the metric theory for the region around a single massive body (like the sun) were given in section 7.5.2. The problem now is to calculate the geodesics, which will be the paths taken by light near such a body. The resulting differential equations turn out to be much like Newton's equations of motion with an additional term bringing in Einstein's gravitational effects. (See Lambourne and Weinberg.) The importance of this additional term was apparent in section 6.6.3 where we saw that it is responsible for the rotation of planetary orbits.

In his 1916, paper Einstein gives the light bending angle as $\theta = 1.75''/\Delta$ where Δ is the closest approach to the sun's center measured in sun radii. (Actually, Einstein gives 1.7 rather than the more accurate 1.75.) We now have a very clear difference for the case of light just grazing the sun ($\Delta = 1$):

classical Newtonian theory (Soldner) $\theta = 0.875''$,
relativity theory (Einstein) $\theta = 1.75''$.

Thus the calculations made a very clear prediction: the bending of light is twice as great if Einstein's theory should replace Newton's. These are crucially important calculations, and I choose them as **calculation 36, bending light**. Here was a very clear prediction, a reference to a phenomenon that had never been observed, and so there was no possibility of the theory being manipulated to fit the data. This is the ideal scientific situation. (For a useful discussion of this point see Brush's paper "Prediction and Theory Evaluation: The Case of Light Bending.") But could light bending really be detected?

9.6.3 The Story of a Triumph

The situation shown in figure 9.13 suggests an experiment: check the positions of stars seen close to the sun against their positions when the light from them does not pass close to the sun. But obviously the faint light from stars is completely swamped by the light from the sun itself. However, if the observations could be made during a total eclipse of the sun, everything changes. The apparent position of stars could be measured during that six-minute window and compared with their known positions already recorded by astronomers. So it was that as World War I ended, scientists were able to leave aside national differences and make expeditions to places where a May 29, 1919 eclipse could be fully observed. An expedition to Principe (off the coast of Spanish Guinea) was led by Arthur Eddington and another to Sobral (in northern Brazil) was led by Andrew Crommelin. (This is one of science's great stories, and to read more, I recommend the book by Will, the biography by Pais, the paper by Kennefick, and Eddington's own account in his 1920 book.)

The remote locations, weather worries, and difficulties involved in making the observations mean that we should not expect great accuracy. Nevertheless, analysis of the experimental data shows that

$$\theta = 1.98 \pm 0.16'' \text{ (Sobral data)}, \qquad \theta = 1.61 \pm 0.40'' \text{ (Principe data)}.$$

While the accuracy is not exceptional, these results clearly favor Einstein's theory. Later measurements gave a similar spread in accuracy (see Weinberg's table 8.1) and even a "modern" 1973 expedition to Mauritania did little better. Later experiments with radio waves and other astronomical situations have confirmed that Einstein's theory is indeed the one to use.

So Einstein's theory passed one of its great tests, and the publicity was enormous (see Pais). But what if it had failed? What would Einstein say then? Einstein believed that his theory of relativity had logical consistency and necessity. Establishing such things was a driving force in his life; he said, "What I am really interested in is whether God could have made the world in a different way; that is, whether the necessity of logical simplicity leaves any freedom at all."[21] When asked what he would have said if the observations had not confirmed his theory of light bending, Einstein replied, "Then I would have to pity the dear Lord. The theory is correct anyway."[22]

The bending of light by distant objects is now an important factor in cosmology. It is known as gravitational lensing (see Lambourne section 7.3.2, for example). The deflecting body may be a galaxy or even a cluster of galaxies.

9.6.4 A Sour Note

Einstein was Jewish, so for some people in Germany, he was far from popular, and efforts were made to discredit his work. He was attacked by those supporting the nationalist movement for "Deutsche Physik" or "Aryan Physics." A leading figure in that movement was Philipp Lenard, who won the 1905 Nobel Prize for experiments in photoelectric effects to which I referred in section 9.5.1. Lenard accused Einstein of plagiarism and claimed that he took his light bending results from Soldner, although there is no evidence that Einstein knew of the earlier work. Lenard published a critical paper in the 1921 *Annalen der Physik*. It was a sad time for Einstein and German science, and, of course, he moved to the United States for the rest of his life.

There is another little twist in this story. In 1911—and before he had fully developed his theory of general relativity—Einstein published a paper "On the Influence of Gravitation on the Propagation of Light" in the *Annalen der Physik*. In that paper, he left out some of the relativistic effects and came up with the result $\theta = 0.83"$, that is, with the classical or Newtonian result. There is no mention of Soldner in that paper (nor would there be for the devious Einstein portrayed by Lenard), but he does suggest the experiment in which stars are observed during a solar eclipse. Perhaps luckily for Einstein, a 1912 experiment in Brazil failed because of rain, and a 1914 expedition to observe an eclipse in Crimea was aborted because of war. Perhaps the "dear Lord" was sparing Einstein some embarrassment until his 1916 paper appeared!

Surely there are few other calculations in science which are surrounded by such remarkable scientific and social outcomes. **Calculation 36, bending light** is truly a landmark in science and in its overall impact.

9.7 AN END TO THE STORY?

We have come full circle, from Newton's corpuscles to Young's light waves, on to Maxwell's electromagnetic waves, and back to particles again with Einstein's photons. What does this say in answer to the question: What is light? To go into this more deeply could take another whole book. Today, we still use the wave theory of light (and other waves like radio waves), and the theories presented in sections 9.3 and 9.4 continue to be used and impress with the range of phenomena so described. There are things that are only explained using a quantum approach. It is a question of choosing the appropriate theory to apply. In a popular article published in 1924, Einstein wrote:

The positive result of the Compton experiment proves that radiation behaves as if it consisted of discrete energy projectiles, not only in regard to energy transfer but also in regard to momentum transfer.[23]

Notice Einstein's use of the words "as if" because that is the way many people like to state the position about light: sometimes it behaves as if it is a wave; at other times, it behaves like a stream of particles. (A wonderful attempt to communicate how quantum theory is used to explain everything about light may be found in Richard Feynman's book *QED*.)

Finally, what should we say about the photon? You might enjoy the various articles collected together by Roychoudhuri and Roy under the heading "The Nature of Light: What is a Photon?" But perhaps it is best to finish with the opinion of Einstein himself toward the end of his life in 1951:

All these fifty years of conscious brooding have brought me no nearer to the answer to the question "what are light quanta (photons)?" Nowadays every Tom, Dick, and Harry thinks he knows it, but he is mistaken.[24]

I have a feeling that Dr. Samuel Johnson would have really appreciated that statement!

Chapter 10

BUILDING BLOCKS

*in which we see some of the important calculations used when
trying to determine the nature of matter and
identify the fundamental building blocks.*

Questions about the properties of matter and what happens when we continue to divide things into smaller and smaller pieces go back at least to the ancient Greeks. Already there was a form of what today we call the atomic hypothesis, although not everyone believed there were final, discrete entities forming the world. The ancient position is beautifully summed up in Lucretius's *De Rerum Natura* (*The Poem on Nature*) written around 50 BCE:

> The whole of nature consists of two elements:
> There are material bodies, and there is empty space,
> In which they are situated and through which they move.
> . . .
> The bodies themselves are of two kinds: the particles
> And complex bodies constructed of many of these;
> Which particles are of an invincible hardness
> So that no force can alter or extinguish them.[1]

Lucretius tries to explain a variety of natural phenomena using that atomic hypothesis and even refers to animal reproduction and the need for "immutable matter" in order for the next generation to maintain the appearances of its parents. Today molecular biology and the study of DNA dominate much of our natural science. In fact it would be hard to overestimate the importance of atoms in the modern world. The *Feynman Lectures in Physics*, given in 1962 to students at the Californian Institute of Technology, are a wonderful source of knowledge, insight, and wisdom. In the very first lecture Richard Feynman talks about atoms:

> If, in some cataclysm, all of scientific knowledge were to be destroyed, and only one sentence passed on to the next generations of creatures, what statement would contain the most information in the fewest words? I believe it is the atomic hypothesis (or the atomic fact, or whatever you wish to call it) that all things are made of atoms—little particles that move around in perpetual motion, attracting each other when they are a little distance apart, but repelling upon being squeezed into one another. In that one sentence, you will see, there is an enormous amount of information about the world, if just a little imagination and thinking are applied.[2]

There are two difficulties here. That "just a little imagination and thinking" is not so easy to supply, and we shall see that it is calculations coupled as usual with experiments that are needed. But the obvious first and more fundamental difficulty for most people relates to the very existence of atoms. As Lucretius says in *De Rerum Natura*, "the original particles, although themselves invisible," may give a plausible explanation for properties of matter, but how do we know that the major assumption that atoms exist is a sound one? It is only in more recent times that tools like x-ray diffraction and electron microscopes have become available for exploring matter at fine scales.

The nineteenth century saw a blossoming of science at these fundamental levels. For example, chemists like John Dalton showed the value of the atomic hypothesis in explaining an array of reactions and chemical properties of substances. The spread of thin layers of oil over the surface of water was seen to have a limited extent, and, assuming a monomolecular layer, Lord Rayleigh and others estimated the size of fundamental molecules and atoms. Some of the most productive "imagination and thinking" saw significant advances in the kinetic theory of gases, that model of a gas as an enormous collection of rapidly moving atoms or molecules as first envisaged by Newton and Daniel Bernoulli. James Clerk Maxwell used statistical methods to show that the kinetic theory led directly to Boyle's Law $PV = RT$ governing how gas pressure P and volume V are linked to the temperature T (R is the gas constant). Maxwell also found a "very startling"[3] prediction that the coefficient of viscosity of a gas is independent of its density (in certain ranges). The experimental verification of that prediction was one of many triumphs for the kinetic theory—and for the atomic hypothesis, of course. (See Brush for the history of this work.)

As great as these advances were, and despite the overwhelming weight of evidence in favor of the atomic hypothesis, there were still some respectable scientists (notably Ernst Mach and Wilhelm Ostwald) who found it hard to accept that atoms existed. For them, the evidence was just too remote.

(The history of this part of science is told in many books. I recommend the "popular" book by George Gamow, the book by Holton and Brush, and the definitive articles by Pais and Rechenberg for the physics and extensive references to key papers.)

10.1 DIRECT EVIDENCE FOR THE EXISTENCE OF ATOMS

Atoms and molecules are not visible, and the tests for their existence were just too indirect for many people. What was needed was a critical test that was somehow visible yet closer to molecular effects or mechanisms. The clinching test was provided by Albert Einstein and Jean Perrin. In 1905, along with his wonderful papers on relativity and quantum theory, Einstein published a paper titled "On the Movement of Small Particles Suspended in Stationary Liquids Required by the Molecular-Kinetic Theory of Heat." (This paper and others mentioned below are available in Einstein's collected papers—see bibliography.) The paper opens with the sentence:

> It will be shown in this paper that, according to the molecular-kinetic theory of heat, bodies of microscopically visible size suspended in liquids must, as a result of thermal molecular motions, perform motions of such magnitude that these motions can easily be detected by a microscope.[4]

Here was a physical effect that (it was suggested) could be observed, and Einstein was in no doubt about its vital importance as he explains just what was at stake:

If it is really possible to observe the motion discussed here, along with the laws it is expected to obey, then classical thermodynamics can no longer be viewed as strictly valid even for microscopically distinguishable spaces, and an exact determination of the real sizes of atoms becomes possible. Conversely, if the prediction of this motion were to be proved wrong, this fact would provide a weighty argument against the molecular-kinetic conception of heat.

Einstein could not be sure that the predicted phenomenon was already known as Brownian motion (see section 10.11):

It is possible that the motions to be discussed here are identical with the so-called "Brownian molecular motion"; however, the data available to me on the latter are so imprecise that I could not form a definite opinion on this matter.

In his doctoral thesis, "A New Determination of Molecular Dimension," Einstein considered particles dissolved in a liquid and developed a theory for their diffusion using known laws for osmotic pressure and motion in a viscous medium. In his 1905 paper, he carried over these ideas to large, observable particles suspended in a liquid; in a *tour de force*, Einstein came to section 4, "On the Random Motion of Particles Suspended in a Liquid and Their Relation to Diffusion" and then the big triumph, section 5, "Formula for the Mean Displacement of Suspended Particles: A New Method of Determining the True Size of Atoms." In 1906, Einstein published a paper "On the Theory of Brownian Motion" in which he derives the same results but with a theory more closely related to "the foundations of the molecular theory of heat."[5]

(Details of Einstein's work may be found in many statistical physics textbooks. Pais, chapter 5, "The Reality of Molecules" is an excellent general coverage of the topic, including Einstein's mathematics, and Rigden gives a simpler account of Einstein's 1905 paper. The papers by Bernstein and by Newburgh, Peidle, and Rueckner are also recommended. The latter gives a brief review of the mathematics and alternative derivations of Einstein's results. Feynman's lecture 41 provides a lovely pedagogical approach to Brownian motion showing how the concept of resisted motion and simple averaging lead to Einstein's results.)

10.1.1 Einstein's Prediction for Brownian Motion

The sort of particle movements Einstein refers to are called Brownian motions after the distinguished botanist Robert Brown (1773–1858), who used a microscope to watch tiny pollen grains as they jerked around in water. At first, Brown thought that the seemingly random jumps must be from some life form ("animalcules") propelling itself through the liquid. As a good experimenter, he tried inert substances, like particles of glass and rock, and still saw the same phenomenon. The vital addition made by Einstein was to establish the law Brownian motion is expected to obey. Here are real predictions put forward for testing.

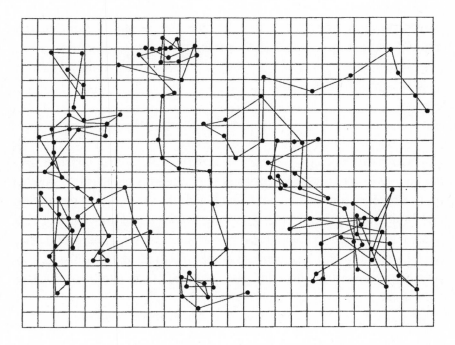

Figure 10.1. Brownian motion of three mastic particles as recorded by Jean Perrin. The positions are given every 30 seconds, and the joining straight lines suggest the zigzag motion involved. *From* Wikipedia*, user MiraiWarren.*

Particles suspended in a fluid jump around because there are fluctuations in the total force imposed by the molecules colliding with them. See figure 10.1. Thus the motion depends on the properties of the fluid, as Einstein expressed in his use of osmotic forces and diffusion. It is obvious from the results shown in figure 10.1 that no formula can be given for the details of the distances traveled by any particular particle. What we can do is to ask about how far a particle undergoing Brownian motion will have moved from its original position and calculate the mean square displacement $<d^2>$ as a function of the elapsed time t. For distance in the x direction, this will be $<x^2>$. This is what Einstein evaluated, with the result that

$$\left\langle x^2 \right\rangle = at \quad \text{so mean distance traveled is} \quad \sqrt{\left\langle x^2 \right\rangle} = \sqrt{a}\,\sqrt{t}. \quad (10.1)$$

The y and z components are similar, and hence the total averaged distance is obtained. Einstein's first prediction is that, on average, the distance moved varies as the square root of the time. (Of course, for uniform, free motion, the distance varies directly as the time.)

Einstein also gave a formula for the proportionality constant a introduced above in equation (10.1):

$$a = \frac{RT}{3\pi\eta\,Nr}. \quad (10.2)$$

In equation (10.2), T is the temperature, R is the gas constant, η is the liquid's coefficient of viscosity, N is Avogadro's number (the number of molecules in a mole of a substance), and r is the radius of the particle undergoing the motion. The gas constant is known and the values of T, η, and r are readily measured. Thus, remarkably, knowing the parameter a would mean knowing Avogadro's number N. (In his doctoral thesis, Einstein had used viscosity data for sugar solutions to estimate N as 2.1×10^{23}; the current value replaces 2.1 with 6.022.)

As scientist and Einstein biographer Abraham Pais puts it:

One never ceases to experience surprise at this result, which seems, as it were, to come out of nowhere: prepare a set of small spheres which are nevertheless huge compared with simple molecules, use a stopwatch and a microscope, and find Avogadro's number.[6]

10.1.2 Experimental Confirmation

Einstein's predictions were confirmed by Jean Perrin (1870–1942) using a series of experiments with a microscope arrangement to record the motion of various different particles. (See Perrin's book for a description and interpretation of his work.) In 1926, Perrin was awarded the Nobel Prize in physics in part for "his work on the discontinuous structure of matter."[7] Molecules had not been observed directly, but the immediate effects of their motions had been confirmed. The link between these molecular motions and the Brownian motion observations could only be made using a theory resulting in calculations like Einstein's. Perrin found the square root of time variation, and his results gave a parameter leading to Avogadro's number $N = 7.15 \times 10^{23}$. Using a theory to link the micro- and macroworlds has allowed us to use measurements of things such as Brownian motion to determine the nature of the atomic or molecular system. (The paper by Newburgh and colleagues shows how Perrin's experiments can be reproduced in a modern laboratory.)

These findings removed the last barriers for the acceptance of the atomic hypothesis. The importance of this work was summed up by Max Born when he wrote that Einstein's theory of Brownian motion did "more than any other work to convince physicists of the reality of atoms and molecules, of the kinetic theory of heat, and of the fundamental part of probability in the natural laws."[8] Surely nobody could fail to add **calculation 37, atoms really do exist** to any list of important calculations.

10.2 LIGHT AND ATOMS

John Dalton (1766–1844) published *A New System of Chemical Philosophy* in 1808, and throughout the nineteenth century the idea of atoms became more and more firmly entrenched in chemistry. The ideas of chemical reactions and valence theory were developed, and the nature of the chemical elements gradually became clear. Dmitri Ivanovich Mendeleev (1834–1907) published *Foundations of Chemistry* and presented his periodic table organizing elements into a system based on their atomic weights, which revealed their chemical properties and identified groups of elements behaving in the same way. However, there was no underlying theory for the nature of the atoms themselves. (Dalton suggested that atoms had a center surrounded by an atmosphere of what he called "caloric," so that neighboring atoms in gases had touching atmospheres.) While the chemists were developing their theories, some physicists were discovering a different—and what proved in some ways to be a more valuable—method of labeling atoms with a characteristic signature. This work was a key step in the lead up to the theory of the atom.

10.2.1 Atoms and Light

It was long known that burning different substances produces light of different colors. It was discovered that in addition to the continuum of wavelengths in radiation (for example, that seen from very hot bodies and black bodies, the subject of Planck's work much later), a heated gas or vapor also emitted light at a set of discrete wavelengths. These are the line emission spectra, and here was a wonderful new tool in science. The collaboration of Gustav Kirchhoff and Robert Bunsen produced the seminal 1860 paper "Chemical Analysis by Observation of the Spectrum." They wrote that

> it is a well-known characteristic of some substances that when placed in a flame certain bright lines appear in their spectrum. These lines open a method of qualitative analysis, extending the field of chemical reactions, and also leading to the solution of problems which have previously been considered inaccessible.[9]

They go on to point out the value of this analysis and an advantage it has over chemical or analytical analysis:

> In spectral analysis, on the other hand, the colored lines stay unchanged and are not influenced by the presence of impurities; the position of the

lines in the spectrum is a chemical characteristic of such fundamental nature as the atomic weight of the substance and can be measured with astronomical exactitude.

Kirchhoff and Bunsen went on to use this analysis to extend the alkali metal group beyond lithium, sodium, and potassium to include cesium and rubidium. Here, they said, was a new tool "to discover the smallest traces of an element on the earth." An example of spectral lines is given in figure 10.2.

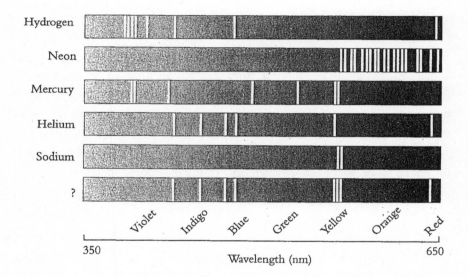

Figure 10.2. Visible spectral lines for a few elements. Each element has its own particular set of lines. The spectrum labeled "?" comes from a gas, and, comparing lines, it can be deduced that it contains helium and sodium. *Reprinted with permission of HarperCollins Publishers Ltd., from Simon Singh, The Big Bang (New York: HarperCollins, 2004).*

It was also in going beyond the confines of Earth that spectral analysis created great interest. In 1814, Joseph von Fraunhofer had detected dark lines in light from the sun, and, in 1859, Kirchhoff had linked these to the emission spectra discussed above. So it came to be known that gases could both emit and absorb light at very particular wavelengths, and this data would allow the responsible elements to be identified. At last, astronomers could find out about the composition of stars and other objects observed in the sky. Husband and wife team William and Margaret Huggins were pioneers in the field. As an example, working in the 1860s, they found dark lines in light from the star Betelgeuse indicating absorption by elements sodium, magnesium, calcium, iron, and bismuth.

Spectral analysis became an essential tool for various scientists, but there remained the question of whether there existed an underlying order and rationale behind the spectral lines.

10.2.2 Bringing Order to the Spectrum

The great step was taken not by someone with a household name like Albert Einstein, Max Planck, or Niels Bohr, but by an obscure Swiss school-master Johann Jakob Balmer (1825–1898). In 1885, Balmer published a paper entitled "Notice Concerning the Spectral Lines of Hydrogen." (It is easy and most worthwhile to read this paper—see bibliography.) He began with the four well-known visible lines in the hydrogen spectrum and came up with a formula for the wavelength λ of the hydrogen lines:

$$\lambda = 3645.6\left(\frac{n^2}{n^2 - 2^2}\right) \quad \text{in units of } 10^{-10}\text{m} \quad \text{and} \quad n = 3, 4, 5, \dots (10.3)$$

This remarkable formula gives wavelength values for a whole series of spectral lines. Balmer gives wavelength values for the first four lines, which he could compare against experimental values found by Anders Jonas Ångström:

Line	n	λ(calculated)	λ(measured)
H_α	3	6562.08	6562.10
H_β	4	4860.8	4860.74
H_γ	5	4340	4340.1
H_δ	6	4101.3	4101.2

In his paper, Balmer also gives comparisons for the next lines using data supplied by Huggins. The fit is almost as good, but later work has shown that Huggins's data contains small errors. Balmer pointed out that the higher spectral lines become more closely spaced and approach the limiting value of 3645.6.

Balmer also speculated about other series of lines obtained by changing $n^2/(n^2-2^2)$ to $n^2/(n^2-p^2)$ and taking other integer values for p. Later it became conventional to write the hydrogen spectra in terms of the inverse of the wavelength (which gives the frequency when multiplied by the speed of light c) with the result that

$$\frac{1}{\lambda} = R\left(\frac{1}{p^2} - \frac{1}{n^2}\right). \tag{10.4}$$

R is the Rydberg constant which has the value 1.09678×10^7 m^{-1}. Choosing a value for p and then taking n to be p plus 1, 2, 3, 4, . . . generates the wavelengths for the whole set of series of lines which are labeled by their discoverer and year of discovery as follows:

$p = 1$ Lyman series (1906–1914)
$p = 2$ Balmer series (1885)
$p = 3$ Paschen series (1908)
$p = 4$ Brackett series (1922)
$p = 5$ Pfund series (1924)

Of course most of these wavelengths refer to radiation not in the visible range.

Equations (10.3) and (10.4) give beautifully simple formulas which describe the hydrogen spectrum with great accuracy. They tell us that there is a definite pattern and mechanism involved in light emission from atoms. Like all good work of this type, they point the way to the future by posing some clear questions: Why do the integers p and n occur in that particular manner? What is the origin of the constants like R at the front of these formulas? Answering these questions led to a revolution in physics as I will describe in the next section.

The work of Balmer and others led to many new results in spectral analysis. For example, in 1896, E. C. Pickering discovered absorption spectral lines in the light from the star ζ Puppis corresponding to the formula

$$\frac{1}{\lambda} = R\left(\frac{1}{(p/2)^2} - \frac{1}{(n/2)^2} \right).$$

(10.5)

In 1912, Fowler saw the same lines emitted from a gas mixture of hydrogen and helium. The link to the existence of helium had to wait for Bohr's theory described in the next section.

It is not clear how Balmer did his calculations to obtain his hydrogen spectrum formula—in his paper he says, "I gradually arrived at a formula."[10] However, his work is enormously important in the development of physics and I add **calculation 38, spectral line patterns** to my list. Certainly Niels Bohr, the founder of the quantum theory of the atom, was in no doubt: "As soon as I saw Balmer's formula the whole thing was immediately clear to me,"[11] he said.

10.3 ATOMS AND THE DAWN OF QUANTUM MECHANICS

Ernest Rutherford (1871–1937) reported on his wonderful experiments and deductions in his 1911 paper "The Scattering of α and β Particles by Matter and the Structure of the Atom." (The book by Ter Haar is an excellent introduction to this part of physics and includes the seminal original papers by Planck, Einstein, Rutherford, Bohr, and others.) Rutherford came to the conclusion that atoms consist of a heavy, positively charged nucleus surrounded by N electrons when the nucleus carries a charge Ne, where $-e$ is the charge of an electron. In the simplest case of hydrogen, $N=1$, and the atom consists of a single proton with one electron accompanying it. Clearly the place to start in understanding the structure of the atom is with hydrogen, and in fact, according to nuclear scientist Victor Weisskopf (1908–2002), "To understand hydrogen is to understand all of physics."[12] Furthermore, the results given by Balmer and his followers show precisely what that understanding must explain.

10.3.1 Enter Bohr

Niels Bohr (1885–1962) was one of the giants of twentieth-century physics. He began his theory of the hydrogen atom by taking the classical mechanics of Newton for the solar system but with the gravitational force replaced by the Coulomb force for the positively charged proton attracting the negatively charged electron. This theory leads to a continuum of possible energies for the electron in orbit around the proton. There is also a major problem: according to electromagnetic theory, an electron moving in such an orbit will radiate energy and so gradually spiral down into the proton. (Bohr's 1913 paper is easy to follow and reproduced in Ter Haar's book.)

Bohr took his cue from Planck's photon theory of radiation and built this into his theory of the atom. Let the electron have energy W, measured as the amount of energy required to remove the electron from its orbit and away to infinity. Building on the discrete photon ideas of Planck, Bohr came to expressions for W and for the diameter $2a$ of the electron orbit:

$$W = \frac{2\pi^2 m e^2 E^2}{n^2 h^2}, \qquad 2a = \frac{n^2 h^2}{2\pi^2 m e E}. \qquad (10.6)$$

In equation (10.6) m is the mass of the electron and $-e$ is its charge; E is the charge of the nucleus so $E = e$ for hydrogen; h is Planck's quantum constant introduced in section 9.5; and n is an integer. Thus in Bohr's theory, the electron only occupies those orbits labeled by the integer n, and it is assumed that those electron orbits are stable.

The orbit with $n = 1$ requires the largest amount of energy to remove the electron from the hydrogen atom leading to what we today call the ground state of the atom. Bohr used the known values of e, m, and h to find

$$2a = 1.1 \times 10^{-8} \text{ cm} \qquad \text{and for } n = 1, \qquad W/e = 13 \text{ volts,}$$

numbers which compared quite well with experimental data.

Bohr could now make the vital step. He assumed that electrons could move between orbits by emitting or absorbing a photon of frequency v such that its energy hv was equal to the difference of the electron energies of those orbits. This led to

$$v = \frac{2\pi^2 me^4}{h^3}\left(\frac{1}{n_2^2} - \frac{1}{n_1^2}\right) \quad \text{or} \quad \frac{1}{\lambda} = \frac{v}{c} = \frac{2\pi^2 me^4}{ch^3}\left(\frac{1}{n_2^2} - \frac{1}{n_1^2}\right). \quad (10.7)$$

This is just the spectral-lines formula in equation (10.4) with an expression for the Rydberg constant given in terms of the fundamental quantities m, e, c, and h. Bohr calculated that $cR = 3.1 \times 10^{15}$ as against the experimental value 3.29×10^{15}. This was taken as excellent agreement given the uncertainties in the input data for his calculation.

Bohr had thus shown that his model for the hydrogen atom, incorporating quantum conditions into classical mechanics, explained the spectral properties of hydrogen so precisely set out by Balmer and his followers. But Bohr could claim another triumph. For ionized helium, there will be one electron orbiting a nucleus with charge $2e$ ($E = 2e$ in equation (10.6)), and then his theory leads directly to the spectral lines observed by Pickering and Fowler (see equation (10.5) and section 10.2.2). Thus their observations showed the existence of helium in its ionized form and fitted perfectly with Bohr's theory.

There were many triumphs for Bohr's theory, but there were also problems, especially moving on to the full helium atom and other systems. A new approach was needed.

10.3.2 Removing the Difficulties
with a Quantum Wave

There is no doubt that Bohr's work is brilliant and deserved the award of the Nobel Prize in 1922, but there is something unsatisfying about the way it cobbles together classical and quantum ideas in an ad hoc sort of way. The resolution to the difficulty takes us into one of the strangest parts of physics, one initiated by a French prince.

Born of a noble family, Louis de Broglie became Prince de Broglie on the death of an elder brother. After various career changes, including time spent in the French army during World War I, de Broglie became a physicist, and the ideas he presented in his doctoral thesis changed physics forever. They led to the award of the Nobel Prize in 1929, and he sets out his intentions clearly in his acceptance speech (see bibliography). After some discussion of the quantum theory of light, he writes:

> This reason alone renders it necessary in the case of light to introduce simultaneously the corpuscle concept and the concept of periodicity.
>
> On the other hand the determination of the stable motions of the electrons in the atom involves whole numbers, and so far the only phenomena in which whole numbers were involved in physics were those of interference and of eigenvibrations.[13]

By this de Broglie means the harmonics found on a vibrating string as in a violin (he was very keen on chamber music), which I discuss further in chapter 12. In this case, only very particular wave configurations occur. De Broglie goes on:

> This suggested the idea to me that electrons themselves could not be represented as simple corpuscles either, but that a periodicity had also to be assigned to them too.
>
> I thus arrived at the following overall concept, which guided my studies: for both matter and radiation, light in particular, it is necessary to introduce the corpuscle concept and the wave concept at the same time. In other words the existence of corpuscles accompanied by waves has to be assumed in all cases.

Here, in de Broglie's final sentence, was the revolutionary step: we must associate a wave with an electron. An electron with momentum p has a wave associated with it with wavelength equal to h/p. Soon, experiments with electrons fired into crystals revealed this wave nature by showing wave-like diffraction effects.

De Broglie's wave "eigenvibrations" could also be found in the fitting of different waves around Bohr's atomic orbits. Here was an explanation of the fact that only very particular orbits occur in the atom; those orbits must be just right for fitting in a series of waves.

The origin of other waves in physics was clear and led to equations which embodied their properties and showed how they propagated. De Broglie's ideas about particle waves were taken on by the Austrian physicist Erwin Schrödinger (1887–1961) and incorporated into the equation which bears his name. Finally, the classical equations of Newtonian mechanics were replaced by the quantum mechanics of Schrödinger's equation.

The wave associated with an electron is conventionally denoted by ψ, and for the electron in a hydrogen atom, the Schrödinger equation takes the form

$$-\frac{\hbar^2}{2m}\nabla^2\psi + V\psi = E\psi$$

(10.8)

where $\hbar = h/2\pi$

and in Cartesian coordinates $\nabla^2 = \frac{\partial^2}{\partial x^2} + \frac{\partial^2}{\partial y^2} + \frac{\partial^2}{\partial z^2}$.

The potential V is simply $-e^2/r$ for the hydrogen atom, and E is the electron energy. Now for the great discovery: solving the Schrödinger equation for the hydrogen atom and imposing the condition that ψ is finite and tends to zero infinitely far from the proton leads to three wonderful results.

First, there is only a discrete set of values E_n for the energy E that accompanies a suitable finite form ψ_n for ψ (x, y, z) that can represent an electron bound to the proton to make the hydrogen atom. This is what is behind Bohr's discovery that only certain electron orbits should be chosen. The form of ψ can be interpreted as representing a standing wave in the space around the proton; these are de Broglie's waves.

Second, those energies E_n are called the eigenvalues, or eigenenergies, and they are labeled by an integer n so that

$$E_n = -\frac{2\pi^2 m e^4}{h^2 n^2}, \qquad n = 1, 2, 3, 4, \ldots$$

(10.9)

These are exactly the energies given by Bohr in his theory of the hydrogen atom. (The minus sign indicates that these are binding energies, and energy must be added to remove an electron from an orbit.)

Third, the electron can only be in one of those wave-states (or eigenstates), and moving between them gives the emission or absorption of energy as a photon exactly as Balmer and others require. The hydrogen spectrum is now explained with the various series of lines linked to the quantum numbers n as shown in figure 10.3.

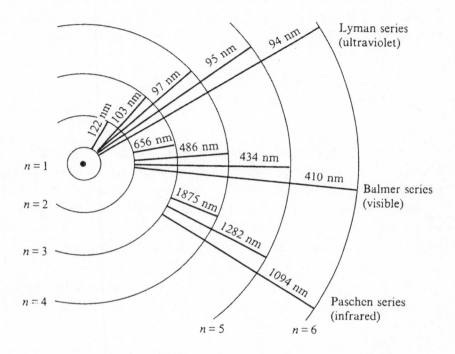

Figure 10.3. Transitions between quantum states showing how the various spectral series are generated. *From Wikimedia Commons, user OrangeDog.*

This is one of the very greatest results in physics, and it goes on my list as **calculation 39, the new mechanics explain atoms**. It paves the way for the whole of atomic physics and beyond that, to modern particle physics.

In general, it is not easy to solve the Schrödinger equation for more complex, multi-electron systems, and approximate methods had to be developed. However, the basic theory was now available for understanding the properties of atoms and molecules, and its validity was proven by the results of the first simple calculations. (The recent book by Fayer uses pictures and largely nonmathematical arguments to show how quantum theory is used for atoms and to understand how molecules may be constructed.)

Another case that is simple but still impressive involves the deuterium spectrum. Deuterium is "heavy hydrogen," an isotope of hydrogen with the nucleus consisting of the proton and also a neutron tightly bound to it. The nucleus now has mass (very close to) $2M$, where M is the proton mass. Now the m in equations (10.8) and (10.9) is really the reduced mass,

$$
m = \frac{m_e M_n}{m_e + M_n} = m_e \left[\frac{1}{1 + \left(m_e / M_n \right)} \right],
$$

where m_e is the mass of the electron and M_n is the mass of the nucleus. Because the electron mass is so small (the proton mass is about 1,836 times larger), the difference between m and m_e is also small. However, calculating the spectrum for hydrogen uses $m = 0.99945 m_e$, while for deuterium it must be $m = 0.99973 m_e$. The outcome is a spectrum for deuterium that is shifted from that of hydrogen by a tiny amount. Yet this tiny difference was still detected in experiments first conducted in 1931 and so deuterium was discovered (see the article by Clark and Reader for the fascinating story).

10.3.3 Quantum Physics

After more than two hundred years, a challenge was made to the fundamentals of Newtonian mechanics, and now we accept that quantum mechanics must be used to properly describe the microscopic, atomic, and nuclear worlds. The price to be paid is the introduction of the wave function ψ and the ensuing struggle to interpret its role in the theory. The probability interpretation (where ψ gives the probability of finding the electron at a particular location) championed by Max Born is widely accepted. (There is an enormous literature on this subject; one of my favorites is the book by Wallace.)

10.4 CHADWICK DISCOVERS THE NEUTRON

The early part of the twentieth century was an exciting time for physicists as they struggled to find the fundamental building blocks for matter. The research of Rutherford and others led to the model of the atom considered in the previous section with a heavy, positively charged nucleus surrounded by very much lighter, negatively charged electrons. It gradually became clear that while the hydrogen nucleus was a single proton, there were particles other than protons present in the heavier nuclei. These other particles carry no charge; they are neutral. Around 1920, Rutherford suggested there might be some other electron-proton combination, but much more closely bound than is the case for hydrogen.

In this same time period, there was much activity in nuclear physics involving the scattering of alpha particles (helium nuclei) from the nuclei of a range of elements. The experimental data gathered revealed that sometimes new elements were produced accompanied by other particles and electromagnetic radiation in the form of gamma rays, for example. By 1930, it had been shown that bombarding beryllium with alpha particles produced carbon and some other particle or radiation. Irène Curie and her husband Frédéric Joliot discovered that this output radiation could knock loose protons from a material such as paraffin.

The decisive steps were taken by James Chadwick (1891–1974), and they confirmed the existence of the particle now known as the neutron. Chadwick was awarded the Nobel Prize in 1935, and both his address on receiving the award and his 1932 *Nature* letter give a good, readable introduction to his work in his own words (see bibliography).

Chadwick considered the various possibilities for explaining the alpha particle-beryllium results, and in his *Nature* letter, he concluded:

> These results, and others I have obtained in the course of the work, are very difficult to explain on the assumption that the radiation from beryllium is a quantum radiation, if energy and momentum are to be conserved in the collisions. [And it would have been a brave person who challenged those conservation laws.] The difficulties disappear, however,

if it be assumed that the radiation consists of particles of mass 1 and charge 0, or neutrons.[14]

10.4.1 Settling the Argument

The charge-neutral nature of the particle was relatively easy to demonstrate, but to show that it was the nucleus constituent required by nuclear physics meant that its mass had to be determined.

In his Nobel Prize–winning address, Chadwick explained how he had done that using simple ideas from classical mechanics. Here was another case of making a calculation to interpret experimental data and so reach an extremely important conclusion.

Assume that particle 1 with mass m_1 is moving with speed V to collide with particle 2 of mass m_2 as shown in figure 10.4. After the collision, the particles have speeds v_1 and v_2. Applying the conservation of energy and momentum laws (see Pask's chapter 13 or any classical mechanics textbook) shows that the second particle leaves the collision with speed

$$v_2 = \left(\frac{2m_1}{m_1 + m_2} \right) V.$$

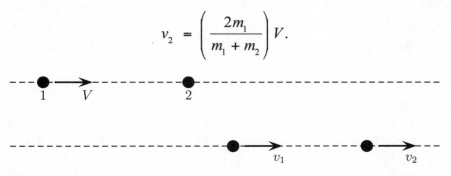

Figure 10.4. Particles 1 and 2 before and after a collision. *Figure created by Annabelle Boag.*

Chadwick used this result for his neutrons (with mass $m_1 = M$) colliding with protons ($m_2 = 1$ and $v_2 = U_p$) and with nitrogen nuclei ($m_2 = 14$ and $v_2 = U_n$) to obtain

$$U_p = \left(\frac{2M}{M+1}\right)V \quad \text{and} \quad U_n = \left(\frac{2M}{M+14}\right)V.$$

These two equations may be combined to give

$$\frac{(M+14)}{(M+1)} = \frac{U_p}{U_n}.$$

If the speeds are measured, this equation may be solved to find the neutron mass M. In his address, Chadwick quotes early results for the approximate speeds as $U_p = 3.7 \times 10^9$ cm/sec and $U_n = 4.7 \times 10^8$ cm/sec, and using those numbers leads to $M = 0.9$. This first effort showed that the neutron mass is of the same order of magnitude as the proton mass; later experiments have given the neutron mass as 1.0014 times the proton mass. (This excess mass over the proton will be important when we return to the neutron story in section 11.3.) After reviewing the experimental and related theoretical work, Chadwick, in his 1932 *Nature* letter, concluded that "Up to the present, all the evidence is in favor of the neutron."[15] Chadwick's announcement of the discovery of the neutron came just seven weeks after the work on the deuterium spectrum described in section 10.3.2.

Thus **calculation 40, discovering the mass of the neutron** proved to be of Nobel Prize–winning importance. The discoverer of the neutron, James Chadwick, went on to be a major figure in nuclear physics and a member of the team developing atomic weapons at Los Alamos. He received many honors and was knighted in 1945.

10.5 THE TRIUMPH OF THE ATOMIC HYPOTHESIS

With the discovery of the neutron in 1932, the list of particles forming atoms (electrons, protons, and neutrons) was complete. The atomic hypothesis was now generally accepted, and the structure of the atoms was revealed, even though it did require a leap into a new and extremely strange aspect of physics. Of course, there were complications and refinements that had to be addressed, but basically the world of atoms and molecules was now established by that powerful combination of theory and experiment. It seemed that Lucretius had got it right:

> The bodies themselves are of two kinds: the particles
> And complex bodies constructed of many of these
> Which particles are of an invincible hardness
> So that no force can alter or extinguish them.[16]

But before too long, and as we will see in the next chapter, doubts emerged about the properties so eloquently described in the last two lines.

It has not been easy to leave out other calculations involving atoms, particularly for crystals and other areas of solid state physics and chemistry. There are times when the limit of fifty calculations seems quite unfair.

Chapter 11

NUCLEAR AND PARTICLE PHYSICS

*in which we see how properties of particles were explored
using the theory-experiment combination and how the
physics of the nucleus solved old mysteries and gave
mankind some terrifying options.*

Atoms were discovered to consist of negatively charged electrons orbiting a positively charged nucleus built from protons and neutrons. The force holding atoms together is the electric, or Coulomb, force between the electrons and the positively charged protons. The Schrödinger equation allows properties of atoms to be calculated, and theory and experiment are in agreement. The theory can then be applied in the realm of molecules. However, going in the opposite direction—down into the nucleus of the atom—opens up a whole new world of physics. Experiments began to reveal strange properties of nuclei and to cast doubt on the comfortable idea that electrons, protons, and neutrons are the only contenders in the structure of the world at its most fundamental level. Also it was found that not just light, electromagnetic radiation, or photons, reached the earth from space; there are also mysterious "cosmic rays" which had to be fitted into the scheme of things.

Although the theory of atomic physics is impressive, there too lurked some tricky problems. In particular, Einstein's theory of relativity stood apart from atomic theory, and Schrödinger's equation seemed to be based on the concepts of classical mechanics rather than the relativistic approach. Relativity treats space and time on a more equal footing than is found in classical mechanics; how should that show up in quantum theory? Einstein had shown that there is a certain equivalence between mass and energy summarized in his famous $E = mc^2$ equation; what are the implications of that in the atomic and subatomic worlds?

The development of those experimental and theoretical challenges led to an undreamed of expansion in the nature of physics and the number of objects to be observed and accounted for. I have chosen six calculations as examples of the steps into modern physics.

11.1 A STRANGE MAN OPENS UP
A WEIRD WORLD

We are now about to enter the weird and wonderful world of quantum field theory. Some people find the concepts mind-blowing, and the mathematics is difficult to follow. Nevertheless, I want to include this section because it leads to one of the most stunning examples of agreement between theory and experiment at an incredible level of accuracy. And I think the main parts of the story can be followed without worrying too much about the mathematics.

The first calculation to be discussed in this chapter tells us about a property of one of the fundamental building blocks, the electron. It also introduces you to a pictorial approach to the organization of calculations.

11.1.1 Enter Paul Dirac

Graham Farmelo's recent biography of Paul Dirac (1902–1984) is called *The Strangest Man*, and perhaps it took such a man to create theories that lead to some of the strangest of all results in physics. Dirac began as an electrical engineer, but lacking a suitable job, he moved to Cambridge University to begin a career in physics. He was legendary for being concise and careful with words. If during question time at a lecture, someone said that they did not understand how the lecturer got to, say, equation six, expecting some words of clarification, they would be met with silence from Dirac who, when pushed by the chairman, would say something like, "oh, I thought that was just a statement of ignorance, not a question."

It is no surprise then that Dirac's 1930 book *The Principles of Quantum Mechanics* is a model of precise writing and logical development of ideas. Recall that Einstein showed how to recast classical mechanics so that Newton's approach was made to accommodate his ideas in the special theory of relativity. Dirac did a similar thing for quantum mechanics, and the Schrödinger equation is replaced by the Dirac equation, which incorporates both quantum ideas and special relativity. Dirac's work on the basic principles of quantum mechanics is one of the pinnacles of theoretical physics. He was awarded the Nobel Prize in 1933 (shared with Erwin Schrödinger), and his acceptance speech (see bibliography) is a fine example of his economical writing style and mastery of physical concepts.

I will introduce three topics emanating from Dirac's ideas and his equation, leading finally to a truly amazing calculation (see section 11.1.4 below). First, the need to meet the requirements of relativity theory compelled Dirac to produce his quantum equation in which Schrödinger's wave function ψ (see section 10.3.2) is replaced by a mathematical object having four components. Clearly, more than the quantum state described by Schrödinger's ψ is involved, and the task was to interpret what the extra components could mean. It turns out that Dirac's equation naturally introduces the intrinsic "spin" of a particle. Already in atomic theory it was found that the individual particles must carry a unit of angular momentum. And since it is natural to think of a particle as some sort of spinning object, the idea of a particle spin came into quantum physics. Dirac showed that it was intrinsic to the full theory and not some added-on artifact.

Secondly, there were extra solutions in Dirac's equation that had strange properties and seemed to imply particles in negative energy states. Again, some interpretation is needed. It turns out that Dirac had found that a particle also has an "antiparticle" with the same mass as the original particle, but with some properties reversed. Thus there is a positively charged antiparticle, known as the positron, as well as the negatively charged electron. The electron and positron have the same mass and spin but opposite charges. The positron was first observed in cosmic rays by Carl D. Anderson in 1933. Dirac had increased the number of elementary particles, since, for example, there would now be a negatively charged antiproton.

Antiparticles are not directly part of our normal everyday matter, but they can be observed in cosmic rays and created in the accelerators used by particle physicists. There is one startling fact that will be important later: a particle and its antiparticle may come together to annihilate one another and change into gamma rays (photons); equally, under the right conditions, radiation may create a particle-antiparticle pair, and virtual pairs may be created and destroyed in the vacuum. Now perhaps you can see just what a strange world Dirac created! The process in which a positron and an electron convert into a pair of gamma rays is the basis for positron emission tomography, which was first explored in the 1950s. Today, it is a technique used widely in medical diagnosis and physiological research. I expect most readers will have seen those amazing maps of brain activity.

Thirdly, Dirac showed that his equation gave the natural way to describe how charged particles interact with radiation and how electrons and photons are coupled together. There were great technical difficulties with this theory (Victor Weisskopf gives a good history), but eventually it turned out to be the most accurate and all-encompassing theory in physics. In *QED: The Strange Theory of Light and Matter*, Richard Feynman gives a masterful account of the quantum electrodynamic theory and how it is used. In his introduction, Feynman writes that "the theory describes *all* the phenomena of the physical world except the gravitational effect . . . and radioactive phenomena, which involve nuclei shifting in their energy levels"[1] (something I turn to later in this chapter). This is quite some claim, and certainly a pioneer like Dirac ought to be forgiven for his strangeness.

11.1.2 The Electron Magnetic Moment

A loop of wire carrying a current generates a magnetic field, and, in some way, the electron's spin property also produces a magnetic effect. Dirac found that the Hamiltonian, the theoretical construct describing how electrons and radiation interact, contains a term involving the electron spin and the magnetic field. He naturally came to the terms that had been introduced in a more ad hoc way by physicists trying to interpret experiments involving electrons in magnetic fields. The electron has a magnetic moment, and it is the product of that with the magnetic field that Dirac found in his new theory. I will write

$$\text{electron magnetic moment} = D \times \text{a constant.}$$

Dirac's equation led naturally to the constant (in terms of the electron's charge and mass). The number D (the Dirac number, as Feynman calls it) is equal to 1 in Dirac's theory.

Developing the description of electron and radiation interactions forced physicists to introduce new elements into the theory and to create what Feynman refers to in the subtitle of his book as "the strange theory of light and matter." In this theory, the number D is no longer 1, and the attempts to compare experimental measurements and calculations of D led to one of the most remarkably accurate, but surely unexpected, results in the whole of science. I give a few technical details in the next section and then present the results in section 11.1.4.

11.1.3 Picturing a Calculation

Among those who built on Dirac's ideas was Richard Feynman (1918–1988), who invented a new way to communicate what was involved in physical processes and how to keep track of calculations. We now know this as the Feynman-diagrams approach. In a space-time diagram, straight lines represent particles, and wavy lines are photons. Thus, when electrons scatter off one another, they do so by exchanging virtual photons as shown in figure 11.1 (a). The diagram is physically suggestive, and it is shorthand for a mathematical calculation that must be carried out to evaluate just what happens in the scattering process. There are rules for translating diagrams into detailed calculations that Feynman gave in his paper "Space-Time Approach to Quantum Electrodynamics." Feynman, along with Julian Schwinger and Sin-itro Tomonaga, was awarded the 1965 Nobel Prize in physics for his work in quantum electrodynamics.

The interaction of charged particles using Coulomb forces, or by considering the electric fields they produce and how they move in an electric field, has been replaced by the quantum picture and the exchange of virtual photons—virtual because energy is not used to create real photons leaving the interaction region; virtual photons pop into existence and then pop out again.

But now we come to the fascinating (and maybe bewildering) part where we recognize that all sorts of other quantum processes may be part of the electron-electron interaction. There can be more than one virtual photon as in figure 11.1 (b). More disconcerting (for many) is the fact that virtual electron-positron pairs may pop in and out of existence as in figure 11.1 (c). A whole chain of increasingly complex processes occurs in the interaction region, and summing up all the contributions to the scattering process is needed to give the total description of what happens when two electrons scatter each other. Fortunately, there is a decrease in the importance of higher terms, and not everything must be calculated to give accurate scattering results.

The interaction of an electron with a magnetic field is also represented in terms of a photon as shown in figure 11.1 (d). Evaluating the implied calculation using Feynman's rules gives Dirac's result (so $D = 1$). But now we must remember that other quantum processes may occur, and some possibilities are shown in figures 11.1 (e)–(g). (Feynman's *QED* book, chapter 3, is a good place to start reading about this, and he shows lots of examples of possible processes.) These diagrams say that a series of calculations must be done which will give

$$D = 1 + C_1\alpha + C_2\alpha^2 + C_3\alpha^3 + C_4\alpha^4 \,... \tag{11.1}$$

In equation (11.1), α is the fine structure constant, which is very close to 1/137. The terms in equation (11.1) become increasingly complex (the C_n requires a more involved calculation) as they correspond to more complex Feynman diagrams, but they become of less importance because the α^n multipliers rapidly decrease.

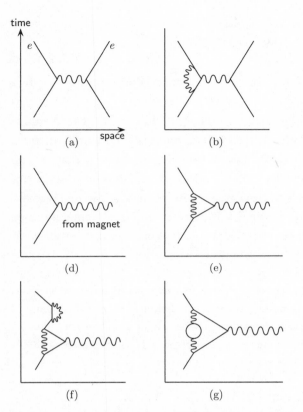
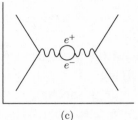

Figure 11.1. Feynman diagrams for particles (straight lines) and photons (wavy lines) in the space-time picture. Diagrams (a)–(c) show electron-electron scattering with a virtual electron-positron pair occurring in (c). Diagrams (d)–(g) indicate processes that must be considered when calculating the magnetic moment of the electron. *Figure created by Annabelle Boag.*

11.1.4 Results for the Electron Magnetic Moment

The calculation of D has been carried out with increasing accuracy over the years, especially as computer mathematical manipulations became available, and the latest (according to Gabrielse, see bibliography) involved the evaluation of almost 14,000 integrals. There has been a corresponding increase in the accuracy of the experimental results, and, according to Gabrielse's 2013 report, the comparison gives

D (calculated) = 1.001 159 652 181 78(77),
D (measured) = 1.001 159 652 180 73(28).

The figures in brackets give the estimated uncertainty in the final given figures.

The stunning agreement between theory and experiment leads me to take **calculation 41, the electron magnetic moment** as a worthy member of my list of important calculations. To emphasize the accuracy of **calculation 41**, Richard Feynman remarks that "if you were to measure the distance from Los Angeles to New York to this accuracy, it would be exact to the thickness of a human hair."[2] It is almost unbelievable that all those strange processes involving photons and virtual pairs of antiparticles can lead Feynman to write in his *QED* introduction:

> The theory of quantum electrodynamics has now lasted for more than fifty years [Feynman is writing in 1985 but there has been no change], and has been tested more and more accurately over a wider and wider range of conditions. At the present time I can proudly say that there is no significant difference between experiment and theory![3]

Yes, almost unbelievable, but true, and other cases are equally as impressive as the example I have chosen to illustrate the accomplishments of twentieth century physicists. Finally, for those of you struggling to make sense of the physical processes involved, let us go back to Dirac:

> The only object of theoretical physics is to calculate results that can be compared with experiment . . . it is quite unnecessary that any satisfactory description of the whole course of the phenomena should be given.[4]

11.2 ONE MYSTERY AND TWO REVELATIONS

In the late nineteenth century, and for the first part of the twentieth century, scientists discovered radioactivity and its challenges to our picture of the particles making up matter. Here was something quite unexpected that required new experimental techniques and new ways of thinking for its investigation. Ernest (later Lord) Rutherford (1871–1937) made great advances with what was termed α-decay, the event in which a nucleus emits an object later identified as a helium nucleus. Rutherford's experiments on the scattering of α particles led to his theory of the atom.

For some nuclei, there was also β-decay, and by 1902, it was clear that the emitted "β rays," or β particles, were in fact electrons. (There are also β-decay processes in which a positron is emitted.) The problem now was to understand the origin of those electrons and the way in which they were emitted by a nucleus.

(I will say more about nuclear physics in section 11.6. A good modern reference is the book by Dunlap. The historical development of the subject is extremely well covered by Pais. For the topic of this section, the article by Brown and the book by Sutton are recommended as easy reading.)

11.2.1 Energies in β-Decay

The α particles are emitted with particular velocities depending on the type of nucleus involved (see Dunlap's figure 8.1 for example), and it was expected that something similar should be the case for β-decay electrons. However, instead of a single "spectral line," there tended to be a continuous spread of velocities. Sometimes several "lines" were observed, and the situation was confused. Furthermore, there is a process called internal conversion in which a gamma ray is emitted and also causes an electron to leave the atom with a particular fixed velocity. We must remember that this was a time when nuclear experiments were still quite new, and particle detection and measurement techniques were not well established.

By 1914, James Chadwick (whose discovery of the neutron I discussed in the previous chapter) was convinced that the β-decay electron energies fell on a continuous curve rather than a set of discrete values. There was then a lull during the First World War. In 1927, Charles D. Ellis and William A. Wooster performed what were taken to be definitive experiments leading them to write that "we may safely generalize this result for radium E to all β-ray bodies and the long controversy about the origin of the continuous spectrum of β-rays appears to be settled."[5]

While it was true that it was accepted that a stream of electrons with variable energies was produced by β-decay, it was not clear how that process actually produced those electrons and which factors then determined their energies. Much of the problem was due to the fact that although the proton was an accepted particle, it was not clear which, if any, other particles helped to form the nucleus. For example, the so-called PE theory held that there were also combined proton-electron entities inside the nucleus. It was not until 1932 that Chadwick definitively established the existence of the neutron.

11.2.2 The Origin of the Continuous Energy Problem

Rather than follow the intricacies of the history of β-decay (see Pais and Brown), let me jump to the result: there are neutrons in the nucleus, and in the β-decay process, a neutron is converted into a proton and an electron is emitted. Note that the electric charge is conserved since the neutron's charge is neutral, and the electron (negative) and proton (positive) charges balance out. Clearly, the β-decay process only occurs for some neutrons in particular nuclei. However, the free neutron always decays with a half-life of 898 seconds.

To understand the electron energy problem, consider the decay of a stationary neutron as shown in figure 11.2. The electron has speed w, the proton has speed v, and they must travel in directly opposite directions to conserve momentum.

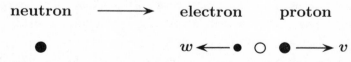

Figure 11.2. Decay of a neutron. *Figure created by Annabelle Boag.*

If the rest masses of the neutron, proton, and electron are M_n, M_p, and M_e, the conservation of energy and momentum laws as expressed in Einstein's relativistic mechanics lead to

$$M_n c^2 = m_p c^2 + m_e c^2,$$

$$0 = m_p v - m_e w,$$

where $\quad m_p = \dfrac{M_p}{\sqrt{1-(v^2/c^2)}} \quad$ and $\quad m_e = \dfrac{M_e}{\sqrt{1-(w^2/c^2)}}.$ (11.2)

These equations may be solved to find the speed of the electron. And now comes the big shock: the calculation reveals that there is just one possible value for w; there is no continuous spectrum for electron energies as the experiments reveal. The single speed predicted is the maximum found for a β-decay electron, but experiments show a continuous set of values as in figure 11.3. Here was an example of a quite straightforward calculation that disagreed with experiment; resolving this difficulty led to a major advance in physics.

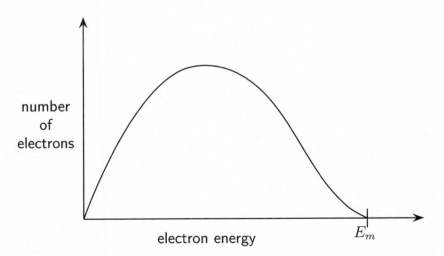

Figure 11.3. Schematic diagram showing electron energies observed in a β-decay process. E_m is the maximum possible electron energy calculated assuming that in β-decay a neutron is converted into a proton and an electron. *Figure created by Annabelle Boag.*

11.2.3 A Technical Aside

For readers wishing to check details for themselves, it is easiest to make the approximation in which an energy mc^2 is approximated as the rest mass energy Mc^2 plus a classical kinetic energy $\frac{1}{2}Mv^2$. Then the conservation laws give

$$(\tfrac{1}{2}) M_p v^2 + (\tfrac{1}{2}) M_e w^2 = B,$$

$$M_p v = M_e w, \tag{11.3}$$

$$\text{where} \qquad B = M_n c^2 - (M_p c^2 + M_e c^2).$$

Notice that B is the energy released because the rest mass of a neutron is greater than the sum of the rest masses of the proton and the electron. (We are of course using Einstein's famous $E = mc^2$ equation.) These equations are easy to solve and

$$w = \sqrt{(2 M_p B)/(M_e^2 + M_p M_e)}.$$

The calculation says this is the only speed that an electron should have after the β-decay of a neutron.

11.2.4 Revelation One: The Resolution of the
Continuous Energy Problem

The phenomenon of β-decay had thrown up a profound problem: apparently the results were in direct conflict with the conservation laws that were absolutely fundamental in physics. It is a sign of the desperation felt at the time that Niels Bohr (of all people) actually contemplated the possibility that energy may not always be conserved. In his 1930 Faraday lecture, Bohr said:

> At the present stage of atomic theory we have no argument, either empirical or theoretical, for upholding the energy principle in the case of β-ray disintegrations, and are even led to complications and difficulties in trying to do so.[6]

To give up such cherished principles would be a drastic thing to do, but the actual solution to the problem proved to be almost as dramatic.

Wolfgang Pauli (1900–1958) was one of the most colorful figures on the quantum-physics scene, and, in typical fashion, he suggested a solution to the β-decay problem in a letter he wrote to colleagues who were to meet in Tübingen in December 1930 (see Pais for full details). The letter is headed "Zürich 4 Dec 1930," and begins, "Dear Radioactive Ladies and Gentlemen."[7] Pauli goes on to propose that there is a "neutron" inside the nucleus, the conservation of energy problem is solved, and

> the continuous β spectrum would then be understandable, assuming that in the β decay together with the electron, in all cases, also a neutron is emitted, in such a way that the sum of the energy of the neutron and of the electron remains constant.

Pauli goes on to say that his solution may sound improbable, "but only who dares wins," and the alternative advice he got was, "the best thing to do is not to talk about it, like the new taxes." He ends his letter by explaining why he cannot come to the conference: "Unfortunately I cannot come personally to Tübingen, because I am necessary here for a ball that will take place in Zürich the night from 6 to 7 December." One of the greatest announcements in physics plays second fiddle to a ball!

I must immediately clarify one point: Pauli's "neutron" is not the same as the one discovered by Chadwick. Pauli said his neutron has small mass and is difficult to detect. It became known as the neutrino—little neutron—a name coined by the Italian physicist Enrico Fermi. Thus the suggestion is that the β-decay process takes the form:

$$\text{neutron} \rightarrow \text{proton} + \text{electron} + \text{neutrino}.$$

Now the energy and momentum are spread over the three particles; there are extra terms to be included in equations (11.2), and solving them leads to a spread of possible electron energies. Thus the calculation of β-decay spectra led to the remarkable prediction of the existence of a totally new kind of fundamental particle. And so I add **calculation 42, why we must have a neutrino** to my list.

The neutrino has very small mass (possibly even zero in some schemes) and has only a minute chance of interacting with other particles. (The neutrino story continues in section 11.7.) Poor Pauli was distressed by this, as the astronomer Fred Hoyle amusingly recounts:

> The astronomer Walter Baade told me that, when he was dining with Pauli one day, Pauli exclaimed, "Today I have done the worst thing for a theoretical physicist. I have invented something that can never be detected experimentally." Baade immediately offered to bet a crate of champagne that the elusive neutrino would one day prove amenable to experimental discovery. Pauli accepted, unwisely failing to specify any time limit, which made it impossible for him to win the bet. Baade collected his crate of champagne (as I can testify, having helped Baade to consume a bottle of it) when, just over twenty years later, in 1953, Cowan and Reines did indeed succeed in detecting Pauli's particle.[8]

Clyde L. Cowan and Frederick Reines were awarded the Nobel Prize for the success of the experiments in what they called their "Project Poltergeist." **Calculation 42, why we must have a neutrino**, not only led to the discovery of one of the weirdest of all particles, it also led to a poem by John Updike: The poem is called "Cosmic Gall" (see Updike's *Collected Poems*), and it celebrates the fact that the weird neutrino can pass right through the earth (and stallions and lovers, Updike notes) without any deviations at all.

11.2.5 Revelation Two: A Left-Handed Universe

The full quantum theory for β-decay was given by Enrico Fermi (1901–1954) and is now known as the theory of weak interactions (a thousand times weaker than electromagnetic interactions). This theory eventually led to a very big surprise: the supposed universal left-right symmetry of the universe breaks down in weak interaction processes like β-decay (technically called the nonconservation of parity). Look at a right hand in a mirror and you see a left hand. It was believed that everything in the physical world had its mirror equivalent. However, in 1956, a paper by T.

D. Lee and C. N. Yang gave a theory for weak interactions in which that left-right symmetry is broken, and, in 1958, their predictions were verified experimentally by C.-S. Wu. Pauli had offered to bet a sum of money that parity would be conserved, but luckily for him, this time nobody took up the challenge! Discussion of left-right, or inversion, symmetry is now prevalent in many fields, and the aptly named book *Right Hand, Left Hand* by McManus is an entertaining and broad introduction to the subject.

11.3 COSMIC RAYS AND A TEST FOR EINSTEIN

I mentioned earlier that Einstein's special theory of relativity is an essential background for the topics in this chapter, especially the relationship between energy and mass. Einstein's theory also makes predictions about time and reference frames that appear quite strange on a first encounter. The observation of particles originating in cosmic rays allowed a test of the theory.

We are used to traveling in trains and airplanes and finding life goes on just the same as when we were at rest on the earth. Galileo used the example of life in the closed cabin of a moving ship. These ideas were built into Newton's mechanics and in his *Principia*, where he writes:

> The motion of bodies included in a given space are the same amongst themselves, whether that space is at rest, or moves uniformly forward in a right line without any circular motion.[9]

Today, we speak about inertial frames of reference. Newton gave the example of a sailor walking along the deck of a uniformly moving ship to explain how speeds are changed depending on the objects to which we refer them (the ship or the earth in the sailor example). Newton's example tells us how to switch between frames of reference, and this established the basis for classical mechanics ever after. Things changed after about two hundred years when Einstein came on the scene.

11.3.1 Special Relativity

Einstein also stated that physics did not depend on the chosen inertial frame of reference, and he went further by including electromagnetic effects in that. Maxwell's theory of electromagnetism (see section 9.4) naturally introduces the speed of light, and thus (following Lambourne's presentation—see bibliography) we have:

The first postulate of special relativity:
> The laws of physics can be written in the same form in all inertial frames.

The second postulate of special relativity:
> The speed of light in a vacuum has the constant value, $c = 3 \times 10^8$m sec^{-1}, in all inertial frames.

The second postulate represents a major step in physics, and, under its influence, Newton's rules for moving between inertial frames must be modified. This makes the change from classical mechanics to relativistic mechanics.

Suppose we have a given inertial frame of reference S and a second inertial frame S' which is moving with speed V in the x direction as shown in figure 11.1. (For simplicity, I keep to the xy-plane—Lambourne gives the full details.) Relativity theory tells us that the change from the coordinates in frame S to those in frame S' is made using a set of formulas known as the Lorentz transformations, and they include the way in which times are specified:

$$x' = \gamma(V)(x - Vt), \qquad y' = y,$$
$$t' = \gamma(V)(t - Vx/c^2), \qquad\qquad (11.4)$$

$$\text{where} \quad \gamma(V) = \frac{1}{\sqrt{1 - V^2/c^2}}.$$

Letting c become infinitely large, so that $\gamma(V) = 1$, gives transformation formulas that Newton would agree with. The inverse transformations tell us how to go from frame S to frame S', that is from x' and t' to x and t.

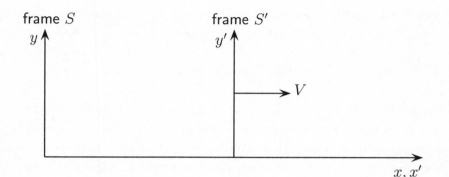

Figure 11.4. A frame of reference S' is in uniform motion with speed V in the x direction relative to the inertial frame S. *Figure created by Annabelle Boag.*

Using the Lorentz transformations makes many changes to Newtonian science which are now well verified. One particular change is perplexing and revolutionary for many people; it concerns time intervals.

11.3.2 Calculating a Time Dilation

Suppose two events occurring at the same place in frame S' have a time interval $\Delta \tau$, then for someone in frame S the time interval ΔT is calculated to be

$$\Delta T = \gamma(V)\, \Delta \tau. \tag{11.5}$$

The transformation factor $\gamma(V)$ defined in equation (11.4) is always bigger than 1 (becoming exactly 1 only for $V = 0$ or when the speed of light c is taken as infinite). The observed time interval ΔT is bigger than the interval $\Delta \tau$.

If we turn the above statement around and say $\Delta \tau$ is smaller than ΔT, we can use the result to state the popular conclusion: moving clocks run slower than those at rest. This is known as time dilation. It is also popularly illustrated in the twins paradox: if one twin goes on a long journey into space on a fast moving rocket, she will be younger than her sister when the two meet up again on her return. Surely few calculations can have led to such a bewildering conclusion! Of course, it is that ratio V/c that is important, and for any significant effect to occur, the rocket's speed V must approach the speed of light c or the transformation factor $\gamma(V)$ must differ only minutely from 1.

11.3.3 The Reality of Time Dilation

There is no doubt that the calculation leading to the time dilation formula is correct, but can we be sure about its meaning? Remember that the postulates of relativity mean that all physical processes must satisfy the theory. Conceptually, it is not easy to appreciate the workings of special relativity. (For the matter of time dilation, readers will find a careful, and largely non-mathematical, discussion in chapter 13 of David Mermin's book *It's About Time*. Mermin assures us that time dilation is real and that there is no intellectual trick in the theory.

Time dilation has now been measured many times (the *Wikipedia* article *Time Dilation* gives a string of references). Using modern atomic clocks and particle accelerators confirms the theory to high accuracy (see the recent papers by Reinhardt and Saarthoff, for example).

The earliest relevant experiments were those on the decay of muons (or mesotrons, as they were called in the early days of this research). Muons occur in cosmic rays and may be detected as they move through the earth's atmosphere at very high speeds. It is worth noting that for great speeds, say 90, 95, and 98 percent the speed of light, we calculate

$$\gamma(V = 0.9c) = 2.29, \quad \gamma(V = 0.95c) = 3.20, \quad \gamma(V = 0.98c) = 5.03.$$

Bruno Rossi observed muons at various heights on Mount Evans in the United States, and his general article "On the Decay of Mesotrons" gives an entertaining description of his exploits and an introduction to the importance of time dilation for his measurements. His detailed results were published in a 1940 paper. An experiment performed by Arthur Greenberg and his colleagues measured the lifetime of moving charged pions and found time dilation confirmed to an accuracy of 0.4 percent. Einstein set out a fundamental part of physics in his special theory of relativity, and testing it was of the utmost importance; **calculation 43, decays and time dilation** must be on my list of important calculations.

Although it might seem that relativistic effects are only to be found in the domain of specialized physical experiments, it should be noted that the global positioning system on which we rely more and more today must take account of those effects for its accurate operation (see the review by Ashby).

11.4 PARTICLE MADNESS

Early in the twentieth century, the atomic world seemed simple; there were electrons (e⁻) orbiting a nucleus of protons (p), and changes in the electron orbit followed photon emission or absorption. Then came the discovery of the neutron (n) as a component of the nucleus. Even today, that picture (a world of electrons, protons, neutrons, and photons) is sufficient to explain the properties of atoms, molecules, solids, and so on. (Michael Fayer beautifully shows how in his 2010 book titled *Absolutely Small: How Quantum Theory Explains Our Everyday World.*)

But refining the theory and probing deeper revealed a richer (potentially more confusing) world. Refining the quantum theory led Dirac to the positron (e⁺) and the idea that every particle also has an antiparticle. Then the decay of a neutron was explained only with the help of the neutrino (v). Cosmic rays included a heavy version of the electron called the muon (μ). Quantum electrodynamics explained the electrical interaction of charged particles in terms of photons leading Hideki Yukawa to suggest in 1935 that protons and neutrons interact by exchanging particles called pi mesons. Pi mesons come in three charged kinds (π^-, π^0, and π^+), and unlike photons, they have mass (about a seventh of the proton mass). Later, heavier K mesons were added to the list.

As particle accelerators and detection devices improved, more and more particles were discovered. There seemed to be a zoo of particles, and physicists looked a little like biologists trying to classify their new findings. In fact, Enrico Fermi quipped that "if I could remember the names of these particles, I would have been a botanist."[10] Some groupings were becoming clear; for example, the electron, its heavier versions (the μ and τ particles), their neutrinos (v_e, v_μ, and v_τ), and all the antiparticles form a group of twelve spin one-half particles known as the leptons. Another large collection of particles are called hadrons, and they divide into the baryons (which include the proton and neutron) and the mesons (including the pions, which mediate the nuclear force). What was needed was a classification showing an order in the same way that Mendeleev's periodic table and the quantum theory of the atom brought order into the world of atoms.

11.4.1 Finding Patterns

Particles were obviously labeled by their mass, charge, and intrinsic spin (as discussed in section 11.1.1). Then it was discovered that new "quantum numbers" could also be attached to particles which now had "isospin" and "strangeness" as properties. (Dunlap's table 2.3 gives a summary of these quantum numbers and their conservation properties.) In the 1960s, Murray Gell-Mann and Y. Ne'eman discovered that arranging particles according to their properties could produce various particular patterns, an idea that became known as the eightfold way. If the spin one-half baryons (which includes the proton and neutron) are arranged in a plane according to their isospin and strangeness properties, a hexagonal pattern is obtained. Similarly, for spin three-halves baryons, a triangular pattern emerges (see figure 11.5). Similar hexagonal patterns emerge for the spin zero and spin one mesons. The triangular pattern including the particle called the Ω^- became particularly famous because this particle had yet to be observed when the pattern was first given.

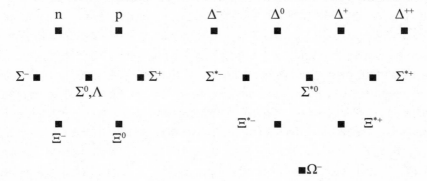

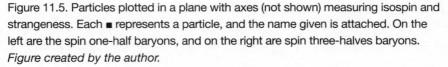

Figure 11.5. Particles plotted in a plane with axes (not shown) measuring isospin and strangeness. Each ■ represents a particle, and the name given is attached. On the left are the spin one-half baryons, and on the right are spin three-halves baryons. *Figure created by the author.*

The pattern organization led to formulas linking particles and their properties such as mass. Patterns are analyzed using a branch of mathematics known as group theory. (For example, all the possible two-dimensional patterns are classified using the seventeen "wallpaper groups," and the possible types of crystals are catalogued using the space groups. In fact I almost included such calculations as a member of my list—going up from 50 calculations would certainly make room for them.) There is a group-theory structure that underpins the patterns of elementary particles and helps us to understand how the patterns form. But most importantly it also points to an underlying simpler structure from which the patterns may be built. That structure predicts a new type of particle called a quark.

11.4.2 Quarks

Elementary particles called quarks were introduced by Gell-Mann. He and George Zweig showed that by using a quark model for mesons and baryons, patterns involving particle masses, lifetimes, and spins could emerge. It now appears that quarks are those elusive unbreakable particles of matter that have been sought ever since the time of the ancient Greeks. (Of course, nobody can be certain—remember, it was once the atom that was taken as the most fundamental unit of matter.)

Current theories involve just six quarks:

up (u) with charge $+\frac{2}{3}$	down (d) with charge $-\frac{1}{3}$	strange (s) with charge $-\frac{1}{3}$
charm (c) with charge $+\frac{2}{3}$	bottom (b) with charge $-\frac{1}{3}$	top (t) with charge $+\frac{2}{3}$

There are also the six antiquarks: u^a, d^a, s^a, c^a, b^a, and t^a. Note that these particles carry fractional charges (when measured in terms of the electron's charge of minus one.) I indicate three quarks x, y, and z, bound together as $|xyz>$. Pairs of quarks and antiquarks form mesons. Here are some examples:

$|uud>$ gives the proton, p	$|udd>$ gives the neutron, d	$|sss>$ gives the Ω^-
$|ud^a>$ gives the π^+	$|u^ad>$ gives the π^-	$|us^a>$ gives the K^+

All of the known baryons and mesons are formed using similar combinations of quarks.

The quarks interact by interchanging particles (bosons) known as gluons, and just as photons come from quantum electrodynamics, gluons come from a branch of quantum theory known as quantum chromodynamics. The theory shows that quarks are confined in their combinations so that no free individual quark can be observed (at least at currently envisaged energies).

This model of fundamental particles goes a long way in explaining the diversity and properties of the old "elementary particles." This is an extremely technical, difficult, and involved theory, but it does appear to be successful. For many people, the crucial test is a calculation of the hadron (baryon and meson) masses in terms of the quark parameters as they are used to fit a small set of data. This was achieved by S. Dürr and his eleven collaborators and reported in a 2008 *Science* paper titled "Ab-Initio Determination of Light Hadron Masses." Their major result is reproduced in figure 11.6. (Note that the masses are measured in terms of the energy unit MeV, or million electron volts, and N denotes a nucleon, which may be a proton or a neutron.) I take this wonderful result as **calculation 44, quarks tell us particle masses**. (This is not a simple area of science; the interested reader may like to consult *The Lightness of Being: Mass, Ether and the Unification of Forces* by Frank Wilczek, Nobel Prize winner in 2004 for his work on quark confinement.) It does appear that we now have an answer to that question of where do all those wretched particles come from: they are the quantum states of a small number of quarks.

It appears now that we have reached the "standard model" for the fundamental units making up the physical world. We have the electron and the other leptons and their neutrinos, and the quarks. (See the short review by Olsen for results about particles comprising more than two or three quarks.) There are also the particles like the photon and the gluons that carry the interaction between leptons and quarks. The search goes on to understand how all of the various particles in the standard model are related to one another and to find an understanding for the basic parameters (like the electron mass) that must be fed into the theory. It has been a long time since Lucretius introduced his "particles of an invincible hardness"[11] and Newton suggested "that God in the beginning form'd Matter in solid, massy, hard, impenetrable, moveable Particles,"[12] and it would be a brave person who declared that the search for the fundamental entities of matter is over, but the many calculations made using the standard model are most persuasive.

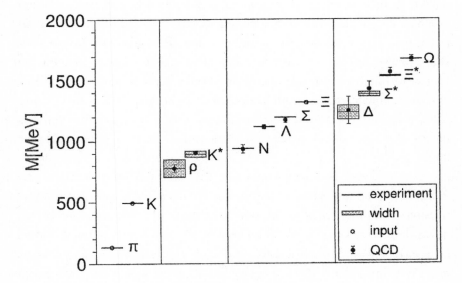

Figure 11.6. Quark model results for hadron masses as published by Dürr and his collaborators. The horizontal bars indicate experimental errors; the vertical bars indicate uncertainties in the calculated results. The π, K, and Ξ masses were used to set quark data. *Reprinted with permission of AAAS, from S. Dürr, et al., "Ab Initio Determination of Light Hadron Masses," Science 322, no. 5905 (2008).*

11.5 INTRODUCING THE NUCLEUS

Before describing the final two calculations for this chapter, it is useful to set out a few facts about nuclei. (To go further, the book by Dunlap is a good reference and the National Research Council (US) book gives an excellent introduction to modern topics in nuclear physics.)

An atom has atomic number Z, meaning that it has Z electrons orbiting a nucleus containing Z protons. The nucleus also contains N neutrons, and $A = Z + N$ is known as the mass number. Protons and neutrons are known collectively as nucleons; the nucleus contains A nucleons. When necessary, an element with the name X will be written as $^N_Z X^A$.

The nucleons interact through the strong nuclear force (which can be related to the exchange of mesons or the interaction of quarks through gluons). The nuclear forces, which tell us how two protons, or two neutrons, or a proton and a neutron interact, are all similar and have a very short range, falling off exponentially. The electric force between the protons is much weaker than the nuclear force, but it is a long-range force and so dominates the p-p interaction at large distances. It is the electric force between protons and electrons that holds the atom together, but it is the nuclear force in competition with the p-p electric force that holds the nucleus together.

The state of the nucleus is given by solving the Schrödinger equation with the appropriate particles and forces included. However, that is a formidable mathematical problem, not particularly simple even when $A = 3$, never mind when ten or even hundreds of nucleons are involved. Physicists have resorted to models like the shell model (which puts nucleons into orbits around an effective center like the electrons in an atom) or the liquid-drop model (which will be used in a later section).

Nuclei vary from hydrogen (with $A = 1$ and $Z = 1$) up to uranium with $A = 238$ and beyond to the transuranic elements known up to $A = 277$. For small numbers Z of protons, the nuclei have roughly the same number of neutrons, but for larger Z, say beyond 25, the trend is for increasingly more neutrons than protons. The number of neutrons in the nucleus for a given Z may vary to give the isotopes, not all of which will be stable; for example, carbon with six protons has isotopes C^A ranging from C^9 to C^{17} although only C^{12} and C^{13} are stable. We have already referred to the α-decay and β-decay processes.

11.5.1 Binding Energies

The next two calculations both refer to the energy contained in nuclei, so some basic information is required. By Einstein's formula, the energy E of a nucleus containing Z protons and N neutrons is given by

$$E(Z,N) = Zm_p c^2 + Nm_n c^2 - B.$$

B is the (positive) binding energy for the nucleus, and it occurs as $-B$ because energy must be supplied to counter the binding and destroy the nucleus. B depends on the nucleons involved, so we must write $B(Z,N)$.

A study over the range of nuclei shows that the binding energy per nucleon, B/A, follows a definite trend: as A increases, B/A increases to reach a maximum around the element iron with $A = 53$, after which it steadily decreases. Using this data, we can see what happens when we have one nucleus with a given A value compared with two nuclei with mass numbers A_1 and A_2 and $A_1 + A_2 = A$. The nuclear reaction will favor the lower energy state, and energy will be given out possibly to electrons, positrons, or photons. This gives us the processes known as fusion and fission:

Fusion: for $A < 53$, two nuclei with mass numbers A_1 and A_2 combine to form a nucleus with mass number $A = A_1 + A_2$ accompanied by energy production.

Fission: for $A > 53$, a nucleus with mass number A breaks up to form two nuclei with mass numbers A_1 and A_2 with $A_1 + A_2 = A$ accompanied by energy production.

An example of fusion given in section 7.3.2 is $F^{19} + H^1 \rightarrow Ne^{20} + h\nu$ where $h\nu$ represents the energy output in a photon.

Energy changes in nuclear reactions are governed by Einstein's $E = mc^2$ equation. In 1932, John Cockcroft and Ernest Walton reported their observations of the reaction

$$Li^7 + p \rightarrow He^4 + He^4 + (8\text{MeV in energy for each } He^4)$$

with the comment that "the evolution of energy on this view is about 16 million electron volts per disintegration, agreeing approximately with that to be expected from the decrease of atomic mass involved in such a disintegration."[13] This is probably the earliest test of this kind and could easily have made my list of important calculations.

(An aside on energies: in atomic physics, a suitable unit of energy is the electron volt eV, which is the energy acquired by an electron in being accelerated through a potential difference of one volt. In terms of other units, $1 eV = 1.602 \times 10^{-19}$ Joules $= 1.602 \times 10^{-12}$ergs. In nuclear physics, the appropriate unit is one million electron volts or MeV.)

These nuclear reactions seem to bring us close to the old transmutation-of-elements dream However, not everyone liked to see it like that; apparently when Lord Rutherford's assistant Frederick Soddy mentioned such things, Rutherford responded that Soddy must not call it transmutation or they would be accused of being alchemists.

11.6 SUNSHINE

Life on Earth depends on the energy, heat, and light radiated by our sun. The sun emits a staggering 3.8×10^{26} watts. The questions of how the sun generates such energy and how long it can last are naturally questions of great importance. The same questions apply to all other stars. The answers take us through a whole set of calculations.

The sun has a mass of around 2×10^{30} kilograms. Gravitational forces have caused the sun to form a dense, hot core with a temperature about ten million degrees Kelvin, reducing to about six thousand degrees Kelvin at the surface. The physics developed late in the nineteenth century and early in the twentieth allowed the properties of such a body to be described in terms of densities and particle motions. There is an internal pressure that counteracts gravitational forces to form a stable structure and prevent further collapsing. A series of calculations allows the internal structure of stars to be understood. (See the book by King for a modern, short introduction to stars.) However, those calculations did not explain how a star produces its enormous energy output.

The situation as it was in 1926 is beautifully set out by Sir Arthur Eddington (1882–1944), one of the great astrophysicists of the twentieth century. (The book by Kilmister gives a short biography and excerpts from Eddington's most important work.) In his *Internal Constitution of Stars*, Eddington explains the problem with the then-current theories and offers a revolutionary solution:

> The energy radiated by the Sun into space amounts to 1.19×10^{41} ergs per year. Its present store of heat energy is as follows
>
> Radiant energy . . . . 2.83×10^{47} ergs
> Translatory energy of atoms and electron. . . . . . 26.9×10^{47} ergs
> Energy of ionization and excitation . . . . $< 26.9 \times 10^{47}$ ergs
>
> This constitutes 47 million years' supply at the most. We do not, however, think that this capital is being used for expenditure; it is being added to rather than exhausted.
>
> It is now generally agreed that the main source of a star's energy is subatomic. There appears to be no escape from this conclusion; but since the hypothesis presents many difficulties when we study the details it is incumbent on us to examine carefully the alternatives.
>
> Formerly the contraction theory of Helmholtz and Kelvin held sway. This supposes that the supply is maintained by the conversion of gravitational energy into heat owing to the gradual contraction of the star. The energy obtainable from contraction is quite inadequate in view of the great age now attributed to the Sun.[14]

We met Lord Kelvin in section 4.3, where his calculation of the age of the earth was discussed. He also worried about the likely age of the sun and the ways in which it could generate energy. In his 1862 article "On the Age of the Sun's Heat," Kelvin concludes:

> It seems, therefore, on the whole most probable that the Sun has not illuminated the Earth for 100,000,000 years, and almost certain that he has not done so for 500,000,000 years. As for the future, we may say, with equal certainty that inhabitants of the Earth cannot continue to enjoy the light and heat essential to their life for many million years longer unless sources now unknown to us are prepared in the general storehouse of creation.[15]

Since the age of the sun is about 4.5 billion years, Eddington was quite right to dismiss Kelvin's theory. He responded to Kelvin's "sources now unknown to us" with the revolutionary idea that the source of the sun's energy is to be found in subatomic processes. We now know that this process is nuclear fusion. We next consider how this conclusion was arrived at and how fusion can operate in the sun since according to Eddington, "No source of energy is of avail unless it liberates energy in the deep interior of the star."[16]

11.6.1 Conditions for Nuclear Fusion

Fusion requires two nuclei to come into contact so that they may join together, or fuse, to form a new nucleus. However, there is one immediate difficulty: nuclei are positively charged, and they repel one another through the long-range Coulomb force. Only when they overcome this repulsion can the strong but short-range nuclear forces take over and pull the constituent nucleons into one bigger, new nucleus. The electrical repulsion is said to produce a Coulomb barrier. The nuclei would need to be traveling at great speed to overcome the Coulomb barrier, and only a few particles in the sun can do that. Was this the end of the nuclear fusion in the sun idea?

The situation was saved by the realization that at the subatomic level we must use quantum rather than classical mechanics. George Gamow developed a theory to show how α particles could overcome a Coulomb barrier; in quantum theory, a particle incident upon a barrier and not possessing enough energy to go over it, can, with some probability, tunnel through it to have the same energy on the inside. This concept was applied to fusion processes in the stars by Robert Atkinson and Fritz Houtermans, and calculations established that fusion was a viable process inside stars. Their 1929 paper paved the way for the final theory, which of course had to come later when more about nuclear physics was known—recall that it was not until 1932 that the neutron was discovered, and the positron followed in 1933.

Houtermans had an interesting and adventurous life (he married four times). There are various versions of a charming story he told about events on the evening after he and Atkinson finished their paper:

That evening, after we had finished our paper, I went for a walk with a pretty girl. As soon as it grew dark the stars came out, one after another, in all their splendor. "Don't they shine beautifully?" cried my companion. But I simply stuck out my chest and said proudly: "I've known since yesterday why it is that they shine."[17]

In some versions of the story, the girl was Charlotte Riefenstahl, who Houtermans married and divorced—twice!

11.6.2 The Grand Solution

In April 1938, George Gamow organized a conference bringing together astrophysicists and nuclear physicists, and the problem of energy generation inside stars was a central topic. Among the nuclear physicists was Hans Bethe (1906–2005). Soon after the conference, Bethe wrote his paper "Energy Production in Stars," giving the solution to the mystery. (Of course there were others involved in this area of nuclear physics such as C. L. Critchfield and C. F. von Weizsäcker.) In 1967, Bethe was awarded the Nobel Prize "for his contribution to the theory of nuclear reactions, especially his discoveries concerning energy production in stars."[18] His 1939 paper is a masterpiece of nuclear physics, with a vast array of information about nuclear processes, including calculations of the details of these processes and the requirements for them to take place.

The star of greatest interest to us is the sun, which is composed largely of hydrogen (over 90 percent) and helium. Therefore, any reactions we find to explain the sun's energy production must involve protons, the hydrogen nuclei. Bethe showed that the basic process involves the conversion of four protons into a He^4 nucleus, which will involve energy production, again calculated using Einstein's $E = mc^2$ equation. The helium nucleus He^4 is stable. To understand the whole process and its energy creation, it is necessary to list the fundamental nuclear processes that must take place. These involve the creation of a deuteron d or H^2, the bound state of a proton (p) and a neutron (n). (Recall that e^+ is the positron, γ is a gamma ray, and ν is a neutrino, and here the electron neutrino is ν_e.) Here is Bethe's scheme:

$p + p \rightarrow d + e^+ + \nu + 0.42 \text{MeV}$ of energy,

$p + d \rightarrow He^3 + \gamma + 5.49 \text{MeV}$,

after two of those steps: $He^3 + He^3 \rightarrow p + p + He^4 + 12.86 \text{MeV}$,

there are two $e^- - e^+$ annihilations: $e^+ + e^- \rightarrow 2\gamma + 1.02 \text{MeV}$.

The whole process may be summarized as

$4p \rightarrow He^4 + 2e^+ + 2\nu +$ energy.

We can now calculate the total energy produced as $2 \times (0.42 + 5.49 + 1.02) + 12.86$ which gives 26.72MeV. About 0.52MeV is taken away by the neutrinos. This may seem like a small amount of energy, but there are an incredibly large number of protons in the sun to take part in the process thus explaining the energy that the sun radiates.

The above process is called hydrogen burning, and Bethe's calculations show that it is a viable process and it does account for the energy produced by the sun. Andrew King calculates that if the sun burnt all its hydrogen, it would last for 10^{11} years, so only a relatively small amount of the sun's hydrogen has been used so far. The sun is essential for life on Earth, and through **calculation 45, why the sun shines**, we understand how the sun supports our planet.

Bethe showed that other nuclear reactions could generate energy in the stars as they are classified in the Hertzsprung-Russell diagram (see King, chapter 1). In particular, Bethe showed that in stars much more massive and hotter than the sun, carbon could act as a sort of catalyst for the protons to helium process and the 26.72MeV energy generation. The chain of reactions involves nitrogen and oxygen and goes like this:

$p + C^{12} \rightarrow N^{13} \rightarrow C^{13} + e^+ + \nu$

$p + C^{13} \rightarrow N^{14}$

$p + N^{14} \rightarrow O^{15} \rightarrow N^{15} + e^+ + \nu$

$p + N^{15} \rightarrow C^{12} + He^4$

Thus the carbon is preserved, which is good because it is not so common in stars. The existence of Bethe's "CNO cycle" is now well established. This theory was discovered independently by Carl von Weizsäcker. Bethe considered many other relevant nuclear reactions in his 1939 paper. (Details of Bethe's paper, along with other related papers mentioned in this section, can be found in the *Source Book* edited by Lang and Gingerich. The book *Fusion: The Energy of the Universe* by McCracken and Stott is a simple introduction to the subject.)

11.6.3 The Story Continues

The fusion process operating in the sun is a clean way to generate energy— there are no nasty radioactive by-products. Furthermore, the burning of one kilogram of hydrogen produces the staggering energy of 6×10^{14} joules whereas burning a kilogram of oil gives us about 4×10^7 joules. That is quite some difference! The obvious question then, is why not use energy generation by nuclear fusion here on Earth? There has been enormous effort to build a fusion reactor, but confining suitable matter to a region with the required temperatures and densities is an incredible challenge. It is not so easy to make a sun on the earth. The technological difficulties have yet to be overcome, although scientists always seem to be optimistic that it only needs a few more years' work! (Interested readers could consult Dunlap's chapter 13 or the book by McCracken and Stott.)

Although the energy-generation mechanism for the sun and other stars found acceptance and there was checking of the nuclear reactions in terrestrial laboratories, a problem arose when considering those pesky neutrinos introduced in section 11.3.4. Eventually, experimental methods were devised for detecting neutrinos, and then the "solar neutrino problem" emerged: the number of neutrinos coming from the sun is only half what was expected. Further experiments and the theory introduced in section 11.5 now suggest that the different neutrinos (v_e, v_μ, and v_τ) associated with the electron, muon, and tau particles are linked in "neutrino oscillations," and the number of electron neutrinos observed is just as it should be. This development also leads to the concept of a very small mass for neutrinos. Recent remarkable experiments have detected neutrinos, giving information about reactions deep in the solar interior. (Consult Dunlap's chapter 17 or the very readable article by Bahcall for more details.) Finally, before we get too diverted, it must be said that neutrinos are now part of astrophysics on an increasing scale (see the article by Halzen and Klein), so Pauli might feel vindicated at last.

11.7 CALCULATIONS THAT FASCINATE AND FRIGHTEN

By 1939, nuclear physicists had reported a discovery that would change the course of human history. They had discovered that heavy nuclei could split into different nuclei with the release of much energy. This process was named fission. The observed results were nothing like α decay; a large nucleus containing many protons splits into two nuclei that both have a large number of protons (in contrast to the two protons in the He^4 nucleus or α particle). The most important nucleus involved in these studies was that of uranium, which contains 92 protons. Fission is observed when neutrons are incident on uranium.

Conclusive experimental results were reported early in 1939 by Otto Hahn and Fritz Strassmann, although in their January *Naturwissenschaften* paper they write, "Now we still have to discuss some newer experiments, which we publish rather hesitantly due to their peculiar results."[19] (This paper and many others in the fission story are conveniently gathered together and discussed by Graetzer and Anderson.) In their February paper, Hahn and Strassmann are much more relaxed writing that "in a rather short time it has been possible to identify numerous new reaction products described above—with considerable certainty, we believe."[20] They talk about the fission by-products as barium, strontium, and yttrium.

As discussed in section 11.6, the theory of a nucleus containing hundreds of strongly interacting component nucleons is extremely difficult, and so physicists struggled to find a theoretical explanation to back up the experimental results. A convincing argument was given by Lise Meitner and Otto Frisch in a January, 1939 *Nature* paper titled "Disintegration of Uranium by Neutrons: A New Type of Nuclear Reaction." (Incidentally, it was in this paper that the word fission was first used in the nuclear context having only been used earlier by biologists when discussing cell division.) It was on a Christmas-holiday walk in the snowy woods of Sweden that Meitner and her nephew Frisch realized that they could use the liquid-drop model of the nucleus developed by Bohr. The analogy between a nucleus and a liquid drop is completed by introducing the right sort of volume terms and surface tension to model the nuclear properties. (As a student in 1906, Bohr had done a project—for which he won a gold medal—on a precision measurement of the surface tension of water by the observation of a regularly vibrating jet, so he was equipped with the classical theory to build the liquid drop model.)

According to Meitner and Frisch: "On account of their close packing and strong energy exchange, the particles in a heavy nucleus would be expected to move in a collective way, which has some resemblance to the movement of a liquid drop. If the movement is made sufficiently violent by adding energy, such a drop may divide itself into two smaller drops."[21] This is illustrated in figure 11.7. Thus the first calculation to explain the fission process began with an analogy that we can picture as in this figure. Meitner and Frisch calculated the energy released in the uranium fission process, and their paper must be ranked as one of the most important in the whole of nuclear physics. (I will say more about Lise Meitner in chapter 13.)

Figure 11.7. Liquid-drop model of a nucleus undergoing fission. *Figure created by Annabelle Boag.*

11.7.1 Details and Implications

Uranium occurs naturally with a mixture of isotopes, mainly U^{238} but with about 0.7 percent of U^{235}. The U^{235} is important because it begins the fission process when neutrons of only comparatively low energies are incident upon it. It becomes an unstable excited U^{236} nucleus, which then decays to complete the fission process to give

$$n + U^{235} \rightarrow Ba^{141} + Kr^{92} + 3n. \tag{11.7}$$

Note that the number of protons remains fixed at 92 (with 36 in the krypton and 56 in the barium), and the original 144 neutrons finish up in the krypton and barium nuclei leaving three neutrons to be emitted. Many processes of this type may occur (see Dunlap chapter 12) with varying numbers of free neutrons to be emitted, with two being a common number.

There are two vitally important points to be made about the uranium fission process. First, it releases a large amount of energy, around 200MeV. To give an idea of what is involved, Hans G. Graetzer and David L. Anderson calculate that the complete fission of a pound of uranium would release more than a million watts of power continuously for a year. Physicists soon realized that here was a possible source of power generation, and, of course, we do have nuclear power stations today.

The second point is the observation that the emitted neutrons were often called prompt neutrons because they may impinge on other uranium nuclei and trigger more fission processes. This we know as a chain reaction, and Enrico Fermi demonstrated experimentally that it can occur (see Graetzer and Anderson for Fermi's paper).

Physicists soon realized that uranium fission could be exploited in a much more sinister way; if the enormous power involved in a chain reaction were released in a short time, the effect would be explosive—the possibility of an atomic bomb became apparent. Remember that this was 1939, and the thought of nuclear weapons for use in the impending World War II was a frightening possibility. (Readers wishing to explore the whole history of this period and the development of nuclear weapons are recommended to consult the Pulitzer Prize–winning book by Richard Rhodes.)

The details of a likely nuclear weapon were set out with great clarity and remarkable accuracy in two memoranda written at the University of Birmingham in England and reached the British-government scientific authorities in March 1940. They were written by Rudolf Peierls, a theoretical physicist and German Jew who fled from Germany before the Nazi era, and Otto Frisch (the nuclear theorist we met above), an Austrian Jew who escaped Nazi Germany in 1939. The Frisch-Peierls memoranda were titled "Memorandum on the properties of a Radioactive 'Super-Bomb'" and "On the Construction of a 'Super-Bomb'; Based on a Nuclear Chain Reaction in Uranium." (You can read these documents—and I recommend them as fine examples of scientific writing—in the book by Robert Serber.)

Albert Einstein wrote a famous letter to President Roosevelt in August 1939 informing him about the possibility of a nuclear bomb and warning him that work on nuclear fission in uranium was being carried out in Germany. (The letter is in the Graetzer-Anderson book.) Einstein suggested that the president should take various actions in response to these developments, and with the receipt of the Peierls-Frisch memoranda in 1941, the way was opened in early 1943 for the Los Alamos laboratories to begin on work to produce a nuclear bomb.

11.7.2 The Los Alamos Primer

Scientists and engineers began arriving at Los Alamos in March 1943. To set the scene and explain what the Los Alamos project was all about, a series of introductory lectures was given in April 1943 by Robert Serber, a theoretical physicist who had worked for Robert Oppenheimer, the head of Los Alamos. Lecture notes were taken by Ed Condon, and a typed version was produced. This became the *Los Alamos Primer*, which Serber published with extended notes and comments in 1992. It is a fascinating and somewhat chilling experience to read the *Primer* and see the calculations that led to one of the turning points in the history of mankind.

The *Primer* opens with a simple statement of intent:

> The object of the project is to produce a practical military weapon in the form of a bomb in which the energy is released by fast neutron chain reaction in one or more of the materials known to show nuclear fission.[22]

Serber points out that the energy release in a nuclear fission process of order 170MeV is about 10^7 times more than the heat energy produced in ordinary chemical or atomic combustion. In order to make the point, he compares the new possibility with an old explosive material and produces the result

$$1 \text{ kg of } U^{235} \approx 20{,}000 \text{ tons of TNT}.$$

This simple but breathtaking comparison would have left no one in doubt about the significance of the project they had all been gathered together to work on.

Serber then gets down to business by evaluating the chain reaction process with a simple calculation:

> Release of the energy in a large scale way is a possibility because of the fact that in each fission process, which requires a neutron to produce it, two neutrons are released. Consider a very great mass of active material so great that no neutrons are lost through the surface and assume the material so pure that no neutrons are lost in other ways than by fission. One neutron released in the mass would become 2 after the first fission, each of these would produce 2 after each had produced fission so in the nth generation of neutrons there would be 2^n neutrons available.
>
> Since in 1 kg of U^{235} there are 5×10^{25} nuclei [in his 1992 book, Serber notes that there is a mistake here; it should be 2.58×10^{24}] it would require about n = 80 generations ($2^{80} \approx 5 \times 10^{25}$) to fish [note the invented verb!] the whole kilogram.

Serber has established what sort of total reaction is required, and now he simply makes the vital and somewhat frightening point:

> While this is going on the energy release is making the material very hot, developing great pressure and tending to cause an explosion.

In fact, for ten percent efficiency, the material would be raised to a temperature of about 10 billion degrees in a millionth of a second. Now, for the first time, comes a hint of the difficulties these bomb designers faced. Some of the neutrons will escape from the material which will now be an expanding hot gas which can ruin the whole process. Serber writes:

> The whole question of whether an effective explosion is made depends on whether the reaction is stopped by this tendency before an appreciable fraction of the active material has fished.

Here is the great difficulty: Can the total reaction occur in a time before the whole thing has expanded out and effectively stopped the explosion? Serber calculates that the reaction must occur in about 5×10^{-8} seconds or "otherwise the material will have blown out enough to stop it."

Serber tests the possibility with a simple order-of-magnitude calculation assuming that neutrons may travel on average (mean free path) about 13 cm between fissions. Here is how he finds the possibility of an explosion occurring:

Now the speed of a 1 MeV neutron [the type released in the fission process] is about 1.4×10^9 cm/sec and the mean free path between fissions is about 13 cm so the mean time between fissions is about 10^{-8} sec. [13 divided by 1.4×10^9] Since only the last few generations will release enough energy to produce much expansion, it is just possible for the reaction to occur to an interesting extent before it is stopped by the spreading of the active material.

The *Primer* goes on to discuss a whole range of technical matters concerning the fission process, including the possibility of using plutonium as well as uranium, as well as methods of detonation. There is also a section headed "Damage" in which the dreadful effects of a nuclear explosion are described and evaluated. However, perhaps the most crucial question concerns the actual size of the bomb. The enrichment process for extracting U^{235} from uranium ore was barely developed at that stage so knowing the amount required could decide the viability of the whole project. There was also the question of the delivery of a nuclear weapon and whether it would be feasible to use an aircraft for bombing enemy territory.

11.7.3 The Size of the Bomb

So far Serber has evaluated the fission process, checked the number of neutrons involved, and calculated the time scales that must be satisfied. Section 10 of the *Primer* is headed "Simplest Estimate of Minimum Size of Bomb." This is a vital calculation to make since the viability of the whole project in terms of a practical weapon rests on the result. Could a nuclear bomb be constructed and delivered to its target?

Serber uses a diffusion-type equation to calculate the number of neutrons involved; there is a generation term (neutrons produced by nuclear fissions) with a time scale of 10^{-8} seconds, as discussed above, and a term describing how neutrons disperse in the medium, which depends on the neutron's mean free path. This theory is used to find the critical radius (assuming a spherical bomb). For a block of material with this critical radius, the neutrons generated are balanced by the number escaping at the boundaries; a larger sphere will explode. The calculation needs boundary conditions, and these must be carefully considered. After some analysis, Serber finds approximations for the critical radius R_c and mass M_c:

$$R_c \approx 9\text{cm}, \qquad M_c \approx 60\text{kg of U}^{235}.$$

To this day, it is hard to believe—and terrifying to contemplate—that a sphere of material just a few centimeters in diameter can cause such a devastating explosion.

Because it is these critical values that are of such vital importance, there is another step that Serber considers. Is there a way to counteract the escape problem for neutrons at the device boundary? Section 11 of the *Primer* begins:

> If we surround the core of active material by a shell of inactive material the shell will reflect some of the neutrons which would otherwise escape. Therefore a smaller quantity of active material will be enough to give rise to an explosion. The surrounding case is called a tamper.[23]

Serber notes that heavy elements (like ordinary uranium U^{238}) should be used for the tamper and gives a mathematical analysis of the role it can play in bomb design. He finally calculates that

> for a normal U tamper the best available calculations give R_c = 6cm and M_c = 15kg of U^{235} while with Au [gold] tamper M_c = 22kg of U^{235}.

There were many unknowns in the whole preliminary analysis, and teams of scientists were needed to work on the project. Serber's conclusion in the *Primer* was that many experiments were needed to measure the neutron and fission properties that went into his calculations. Nevertheless, the calculations that I list as **calculation 46, planning for a bomb** indicate that the project was indeed viable. The horrific results of the August 1945 bombing of Hiroshima and Nagsaki showed just how effective those calculations were.

11.7.4 Uncertainty about Heisenberg: Hero or Poor Calculator?

Albert Einstein, in his letter to President Roosevelt, warned about decisions made in Germany to stop sales of uranium from the Czechoslovakian mines, which invading German armies had taken over. In the Frisch-Peierls memoranda we find another warning:

> If one works on the assumption that Germany is, or will be, in the possession of this weapon, it must be realized that no shelters are available that would be effective and could be used on a large scale. The most effective reply would be counter-threat with a similar bomb. Therefore it seems to us important to start production as soon and as rapidly as possible, even if it is not intended to use the bomb as a means of attack.[24]

Obviously, the possibility of a nuclear-fission bomb was understood in Germany just as it was elsewhere. Many scientists fled from Germany and Nazi persecution, but a number of eminent nuclear physicists and chemists remained. Among them was Werner Heisenberg, one of the founders of quantum theory and the recipient of the Nobel Prize for Physics in 1932. Heisenberg was a leading figure in German efforts to develop nuclear weapons. He was not a Nazi, but he was a patriotic German and hoped for a German victory.

The story of Heisenberg's involvement is complex and murky, and several books have been written on the subject. The central question appears to be: Did Heisenberg deliberately make an error that effectively sabotaged the German nuclear-weapons program? He calculated a critical mass of tons of uranium, rather than the real value of a few kilograms mentioned above. Such a result would immediately signal that the bomb project was not viable. Was that a deliberate error?

However, the evidence seems to point to the fact that Heisenberg made a genuine and major blunder when he did the calculation; he was just not careful enough with the theory and its evaluation. (Interested readers should consult the article by Jonothan Logan and the book by Paul Rose for detailed and meticulously presented arguments.) There are those who have tried to present Heisenberg as some sort of hero for foiling German nuclear-weapons plans, but the fact casts grave doubts on such a story.

The most telling evidence comes from the words of Heisenberg himself. A group of German nuclear scientists was interned near Cambridge, England, at a place called Farm Hall from July to December 1945. The group included Werner Heisenberg, Otto Hahn, Max von Laue (famous for his work on x-rays), C. F. von Weizsäcker, Walter Gerlach, and other notable experts. British intelligence officers recorded the group's conversations. On August 6th, the group heard the news that a nuclear bomb had been dropped on Hiroshima. They were clearly stunned and thought a trick was being played. We have Heisenberg's own words:

> I still don't believe a word about the bomb but I may be wrong. I consider it perfectly possible that they have about ten tons of enriched uranium, but not that they have ten tons of pure U235.[25]

It is clear that Heisenberg still believes that the critical mass is to be measured in tons not kilograms, and he persists with that line over the next few days. He goes on to suggest that the report on the bomb was a fraud:

> All I can suggest is that some dilettante in America who knows very little about it has bluffed them in saying "If you drop this it has the equivalent of 20,000 tons of high explosive" and in reality it doesn't work at all.

Heisenberg was always a confident man and sure of his abilities, so he must have been stung by the chemist Hahn's comment:

If the Americans have a uranium bomb, then you're all second raters. Poor old Heisenberg . . . You're just second-raters and you may as well pack up.

(This is a fascinating part of the history of physics, and I recommend the book introduced by Sir Charles Frank to anyone wishing to read the Farm Hall conversations in detail and learn more about the participants and what became of them after the war. The transcript of the German scientists' conversation immediately after learning about Hiroshima bombing and their debates for days afterward makes for truly gripping reading.)

There are a great many documents and statements related to Heisenberg's war record but amazingly there are still areas of confusion and disagreements about interpretations. Personally, I find the writings of Jonothan Logan and Paul Lawrence Rose most convincing, and I am prepared to believe Rose's final position:

This German mentality of Heisenberg and his friends, fertilized by astounding powers of self-delusion and rationalization, spun the tissue of deception and self-deception that produced the Heisenberg version and the cocoon of fabrication and denial that has blurred the history of Heisenberg's work on the atomic bomb to the present day.[26]

Although Heisenberg did meet his old colleagues, like Niels Bohr, after the war, he never recovered his preeminent position in the world of science.

11.8 TOWARD A FINAL PICTURE OF MATTER

The last one hundred years have seen immense progress in our understanding of the basic particles of matter and the way they move and interact. The standard model represents a concise structure for all matter—except, apparently, for the dark matter discussed in section 7.4! There is also the worry about the nature of dark energy. However, the standard model combined with relativistic concepts helps to explain the origin of the universe and its development after the big bang. But not everything is completed. There is still that nagging question of how to combine general relativity and quantum theory. The standard model itself asks some fundamental questions: Why do we find those particular basis particles, and why do the mass and other parameters take on those particular values? No doubt there are many great calculations still to come.

Chapter 12

METHODS AND MOTION

*in which we return to applications of classical mechanics;
and see how some of the great techniques of applied mathematics
were discovered.*

I have discussed the successes of classical, or Newtonian mechanics, in describing the solar system and how the modifications imposed by relativity and quantum theory have taken science to new triumphs. In this chapter, I return to some of the earliest work in classical mechanics that has led to mathematical techniques of universal importance in both science and mathematics. In the twentieth century, this early work has been combined with new computational techniques to carry out calculations that have changed our views on dynamical systems and their behavior.

This chapter is a little more mathematically detailed than the previous ones, especially for the first two topics, but even skimming over the details should allow everyone to appreciate the results.

12.1 HARMONICS AND
NEW WAYS OF THINKING

The pendulum was of great importance in the early development of mechanics by Galileo, Newton, and Huygens. Its motion was explained theoretically, and it provided a tool for exploring the laws of motion and the effects of gravity. The next step was to deal with continuous systems; so for example, the pendulum (a weight at the end of a light string) would be replaced by a heavy rope or chain hanging down from a suspension point. I return to the motion of such a suspended system in section 12.2.

The problem which received much attention and led to some revolutionary mathematical thinking was that of a vibrating string such as we find in a guitar or other stringed instrument. It may come as a surprise to many people, but it was from studies of vibrating strings that some of the most important of all physical concepts and mathematical descriptions evolved.

Assume that a string has length L, mass ρ per unit length, and is under a force giving it a tension T. The string is stretched along the x-axis and vibrates so that its shape at time t is given by $y(x,t)$. See figure 12.1. The dynamical problem is to find the motion of the string given the initial shape $y(x,0) = y_{in}$ and assuming the string is released from rest. (This is a standard problem found in introductory physics texts and books on waves such as that by Coulson and Jeffrey.) The sound frequencies produced by vibrating strings were considered by Pythagoras and the evaluation of the string's shape was a key problem in the history of mathematical physics.

By necessity, this topic uses some detailed mathematics; in section 12.1.2, I have summarized and interpreted the results and their implications in a general way. (Readers wishing to see more details of the history of this topic may consult Morris Kline's comprehensive history, and check the relevant papers (referred to below) and commentaries on them in the book by Cannon and Dostrovsky and in the sourcebooks edited by Magie and Struik.)

12.1.1 Discovering the Theory of the Vibrating String

In 1636, Marin Mersenne, taking up where Pythagoras left off, we might say, wrote about "Musical Tones Produced by Strings" and gave rules explaining how these tones depended on string length, tension, and so on. A new era began in 1713 when Brook Taylor considered the dynamics of the particles making up a string and used analogies with pendulum motions to show that a stretched string vibrated with a frequency $v_1 = \sqrt{(T/\rho)}/2L$ and had the shape of a sine curve.

Taylor knew that the vibrating string did not always have this shape, but he assumed that after a number of vibrations it would settle down into this proper shape. (Incidentally, this is the Taylor who remains famous for the mathematical Taylor series.)

Both Joseph Sauveur (inventor of the term acoustics) in 1713 and John Wallis in 1741 wrote about other forms that the vibrating string might assume, noting in particular that there could be nodes—points on the string that remained at rest. Examples are shown in figure 12.1.

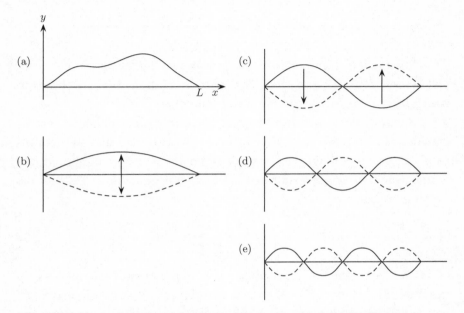

Figure 12.1. (a) General form of vibrating string of length L. (b)–(e) Shape of vibrations corresponding to the first four frequencies for a vibrating string. Note that in (c)–(e) there are nodal points at which the string remains stationary. *Figure created by Annabelle Boag.*

The full mathematical analysis of the vibrating string was largely the work of three of the premier mathematicians of the eighteenth century: Daniel Bernoulli (1700–1782), Jean le Rond d'Alembert (1717–1783), and Leonard Euler (1707–1783). In one first step, d'Alembert used a model of the string as a system of linked particles, applied a limiting process, and thus derived the famous wave equation for the displacement $y(x,t)$:

$$\frac{1}{c^2}\frac{\partial^2 y}{\partial t^2} = \frac{\partial^2 y}{\partial x^2} \qquad \text{where } c = \sqrt{\frac{T}{\rho}} \text{ is the wave speed.} \quad (12.1)$$

D'Alembert gave a solution for this equation in terms of traveling waves (see Coulson and Jeffrey for example). But the solution of greatest interest, the one championed most by Daniel Bernoulli, was the one written as a sum of the string harmonic shapes and frequencies:

$$y(x,t) = \sum_{n=1}^{\infty} a_n \sin\left(\frac{n\pi x}{L}\right)\cos\left(\frac{n\pi ct}{L}\right). \qquad (12.2)$$

This solution corresponds to the string of length L being released from rest in some form $y(x,0) = y_{in}(x)$. The constants a_n in the sum are chosen to match that initial release condition (see below).

For many years there was much controversy surrounding the use of equation (12.2) (see Kline for the intriguing story), and Euler and Joseph-Louis Lagrange battled with Daniel Bernoulli over its validity. On the mathematical front, a question arises when we use equation (12.2) for the initial $t = 0$ case:

$$y(x,0) = y_{in}(x) = \sum_{n=1}^{\infty} a_n \sin\left(\frac{n\pi x}{L}\right). \qquad (12.3)$$

Euler had argued that the initial shape of the string could take a very general form, which included shapes with abrupt changes in slope (that is, continuous curves, but not curves with a continuous derivative). Euler and Lagrange did not accept that such general curves could be matched by a series of sine functions; Daniel Bernoulli claimed that there were an infinite number of constants, a_n, and that allowed all functions to be accommodated. At the heart of this controversy was the definition of a function. It turns out that Bernoulli was correct, but a number of years went by until Joseph Fourier and others provided the background and rigor to settle the question in his favor.

Equations (12.2) and (12.3) give us the solution of the vibrating string problem, and it only remains to find a way to calculate the constants a_n for a given initial displacement of the string. Eventually it was discovered that each constant may be found by carrying out a particular integration:

$$a_n = \frac{2}{L}\int_0^L y_{in}(x)\sin\left(\frac{n\pi x}{L}\right)dx. \qquad (12.4)$$

We now have a complete solution for the vibrating string problem; evaluating equations (12.2)–(12.4) tells us exactly how the string shape $y(x,t)$ evolves from any given input shape $y_{in}(x)$. That is one of the great triumphs of mathematical physics.

12.1.2 Interpretation

A stretched string can vibrate in particular modes or patterns. The nth mode vibrates with a frequency v_n depending on the string parameters T, ρ, and L: $v_n = nc/2L = n\sqrt{(T/\rho)}/2L$; and has a pattern with $n-1$ nodes. The $n=1$ mode is called the fundamental and modes with larger n are the harmonics. Examples are shown in figure 12.1. Mathematically, the modal pattern is described by a sine function, and the vibration has amplitude varying periodically in time according to a cosine function. Thus each term in equation (12.2) represents one of these modes given an initial amplitude a_n.

If the initial shape of the string is exactly in the form of one of those modal patterns or sine functions, then the string will vibrate so that pattern is maintained for all time with its amplitude varying with the modal frequency v_n. There are an infinite number of these modes.

It is most likely that the initial shape of the string does not exactly match one of the modal patterns. In this case, equation (12.2) gives us the remarkable result that the motion of the string will be a mixture of modes simply added together with appropriate amplitudes to match the initial displacement. Here is the way Daniel Bernoulli put it:

> My conclusion is that all sonorous bodies include an infinity of sounds with a corresponding infinity of regular vibrations. . . . We remark that the chord AB [the string, in this case] cannot make vibrations only conforming to the first figure [fundamental] or second [harmonic] or third and so forth to infinity, but that it can make a combination of these vibrations in all possible combinations.[1]

An example will help to appreciate this result, and then I will discuss the wider implications for mathematics and science.

12.1.3 Playing Pizzicato

Some music requires stringed instruments like violins to be played pizzicato, that is they are plucked rather than excited using a bow. The plucked string is the most famous example of the use of the above theory. The problem is illustrated in figure 12.2. The string is initially in a triangular shape with its midpoint raised a distance h above the x axis.

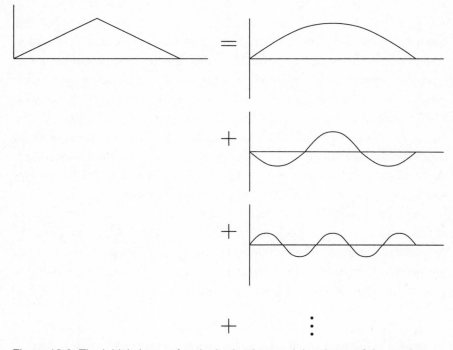

Figure 12.2. The initial shape of a plucked string, and the shape of the modes excited when it is released. (Not to scale; the harmonics have much smaller amplitudes than those shown here.) *Figure created by Annabelle Boag.*

It is easy to find the modal amplitudes using equation (12.4) and

$$a_n = \left(\frac{8h}{n^2\pi^2}\right)\sin\left(n\pi/2\right). \qquad (12.5)$$

Only the odd-numbered modes are excited because the sine term in equation (12.5) is zero whenever n is even. The odd-mode amplitudes oscillate in sign and because of the n^2 term they rapidly decrease in magnitude. In the case of the plucked string we can describe its motion by writing its shape at time t after release as

$$y(x,t) = \left(\frac{8h}{\pi^2}\right)\left\{\sin\left(\frac{\pi x}{L}\right)\cos\left(\frac{\pi ct}{L}\right) - \frac{1}{9}\sin\left(\frac{3\pi x}{L}\right)\cos\left(\frac{3\pi ct}{L}\right) + \frac{1}{25}\sin\left(\frac{5\pi x}{L}\right)\cos\left(\frac{5\pi ct}{L}\right) - \ldots\right\}.$$

This result is shown graphically in figure 12.2. Obviously the excited modes must have the same symmetry as the starting configuration, and that is why those modes with $n = 2, 4, 6, \ldots$ are not present in the string's motion.

We can see that plucking a string at its midpoint strongly excites the fundamental mode; other modes are also excited (in theory, an infinite number of them), but their importance rapidly decreases with increasing mode number. Representing the motion of the string in terms of the modal vibrations gives us a simple way to see which frequencies are generated and just how the harmonics enter to give that attractive composite plucked-string sound.

12.1.4 Going Further

The theory outlined above is often called Fourier theory since it was Joseph Fourier (1768–1830) who fully developed and expanded the ideas encapsulated in equations (12.2)–(12.4). Fourier showed how these same ideas could be used to understand the conduction of heat and described his work in his 1822 book *The Analytical Theory of Heat*. Lord Kelvin (as William Thomson then) was one of the first great enthusiasts for Fourier theory ("Fourier is a mathematical poem"[2] he said), and if you look back to chapter 4, you will see that Kelvin used Fourier's approach in his studies of the age of the earth and in his evaluation of tides. Lord Kelvin recognized the power of Fourier's methods when he and Peter Guthrie Tait wrote in their famous textbook that they must consider

Fourier's Theorem, which is not only one of the most beautiful results of modern analysis, but may be said to furnish an indispensable instrument in the treatment of nearly every recondite question in modern physics.[3]

Kelvin would have been interested to see the growth of Fourier optics in which Fourier's methods are used to describe optical effects and to evaluate imaging instruments. The resolution of image details discussed earlier in section 9.3.2 is beautifully described in terms of spatial frequencies (see the review by Westheimer and the book by Wandell).

The theory has been generalized so that the sine functions used above are replaced by other functions found by solving the differential equation relating to the physical problem under investigation (as sine waves were found by solving the wave equation, equation (12.1), for the motion of a stretched string). If these new functions are $\psi_n(x)$, and suitable conditions are satisfied, the function to be found may be written as

$$f(x) = \sum_{n=1}^{\infty} a_n \psi_n(x). \tag{12.6}$$

All of the ideas discussed above about calculation methods and interpretation of solutions carry over into this new area. Further examples will be given in the next section. (A comprehensive review and history of this area may be found in the book by Elena Prestini.)

This wonderful method becomes quite general: split the calculation into two parts. The first part involves finding the suitable basis functions like sines or $\psi_n(x)$ in general; the second part involves finding the coefficients a_n to match the conditions imposed by a particular problem.

We sometimes speak of the method of superposition, because that is what is happening in equations (12.2) and (12.6) for example; many basis solutions are being superposed to build up any required solution. The building up of solutions using the simplest basis solutions is used for linear systems in general (Feynman may be consulted for an excellent general introduction). The wave equation, equation (12.1), Fourier's heat conduction equation, electromagnetic wave equations, and quantum theory are all linear systems, and I will say more about such things in section 12.3. It is interesting to note that Paul Dirac's superb book *The Principles of Quantum Mechanics* (discussed in section 11.1) opens with a chapter called "The Principle of Superposition."

It seems unlikely at first sight that understanding how a plucked string produces that lovely pizzicato sound could be important, but it does lead to what may well be argued as the most powerful method in mathematics and its applications in science. The calculations made by Daniel Bernoulli and his contemporaries showed the way for physicists to follow ever since. I must certainly put **calculation 47, strings and Fourier's mathematical poem** in my list of significant calculations.

12.2 TABLES OF BESSEL FUNCTIONS

It would be a rare person who used mathematics in scientific problems but who did not come across Bessel functions. However, it is likely that most of those persons know little about Friedrich Wilhelm Bessel (1784–1846) himself.

Bessel was born in Minden, Germany, and went on to become one of the outstanding astronomers of his time. In 1810, at the age of just 25, he was appointed director of the newly founded Königsberg Observatory, and he remained there for the rest of his life despite being offered positions like director of the Berlin Observatory. He made a vast number of observations of stars and processed them to contribute to the catalogues then being assembled. He supplied many of the observational and analysis techniques for that work and published details in 1830 in his great *Tabulae Regiomontanae*. (See the article on astronomical tables by Norberg and Fricke for a comprehensive biography.)

Bessel is remembered for two particularly impressive feats. In 1838, he became the first person to accurately find the distance to a star. Using the parallax method with the diameter of the earth's orbit as base, Bessel found a parallax angle of 0.314 arcseconds for the star 61 Cygni, which translates into a distance of 10.3 light years, a result now known to be in error by less than 10 percent. Finally, astronomers had a grasp of the scale of the cosmos. Secondly, Bessel observed variations in the motion of Sirius which could be interpreted as the gravitational effect of another large body, and, in 1844, he announced that there must be a "dark companion" for the star Sirius. It was not until 1862 that Alvan Clark finally observed the white dwarf that shares its motion with Sirius. Bessel was an outstanding mathematician, and it is for his work on the functions that now carry his name that he is best known today.

12.2.1 What are Bessel Functions?

Many people first meet Bessel functions through the differential equation

$$x^2 \frac{d^2 y}{dx^2} + x \frac{dy}{dx} + (x^2 - n^2)y = 0. \qquad (12.7)$$

This equation, or versions of it, turns up in an enormous number of applications of mathematics to physical problems. I will take x and n to be real, although the theory extends into the complex domain. A solution to this equation is $y = J_n(x)$, which is the Bessel function of the first kind and of order n. (For simplicity I will not consider the other kinds here; in fact $J_n(x)$ is the Bessel function most commonly met with in first applications, usually with integer values $n = 0$ or 1.) Standard methods for solving the differential equation produce series expansions:

$$J_n(x) = \sum_{r=0}^{\infty} \frac{(-1)^r (x/2)^{n+2r}}{r!(n+r)!}$$

$$J_0(x) = 1 - \frac{x^2}{2^2} + \frac{x^4}{2^2 4^2} - \frac{x^6}{2^2 4^2 6^2} - \cdots \qquad (12.8)$$

The graphs of $J_0(x)$ and $J_1(x)$ are shown in figure 12.3. Bessel functions sometimes turn up in integral form such as

$$J_n(x) = (1/2\pi) \int_0^{2\pi} \cos(n\theta - x\sin\theta)\, d\theta$$

$$\text{so} \quad J_0(x) = (1/2\pi) \int_0^{2\pi} \cos(x\sin\theta)\, d\theta. \qquad (12.9)$$

A most useful property of Bessel functions is the recurrence relation:

$$J_{n+1}(x) = \frac{2n}{x} J_n(x) - J_{n-1}(x). \qquad (12.10)$$

Putting $n = 1$ tells us that we can find the value of $J_2(x)$ if we know the vales of $J_0(x)$ and $J_1(x)$, and then we can move on to $J_3(x)$, and so on.

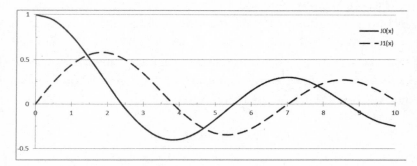

Figure 12.3. Graphs of $J_0(x)$ and $J_1(x)$.

The graphs indicate that these Bessel functions have reducing magnitudes and oscillate in sign, and, in fact, they have an infinite number of zeros. The zeros are not equally spaced (as they are for sines and cosines) although they do tend toward that property for very large values of x as can be deduced from the asymptotic, large x approximation:

$$\text{for large } x: \quad J_n(x) \sim \sqrt{\frac{2}{\pi x}} \cos\left(x - \frac{\pi}{4} - \frac{n\pi}{2}\right).$$

Bessel made the first systematic study of the functions now bearing his name in 1824, and since then, the literature concerning them has become enormous. (G. N. Watson's comprehensive *Treatise* on the subject runs to over 800 pages.)

12.2.2 Using Bessel Functions

As the use of mathematics in science developed, the need for Bessel functions arose. (The history is covered by Watson in his chapter 1 and in the paper by Dutka.) A few examples will set the scene.

It was natural for early researchers to pass from a study of the motion of the simple pendulum to the oscillations observed in a hanging heavy rope or chain. What is the displacement y of the rope from the vertical as a function of the distance x along the rope and the time t? See figure 12.4 This problem was solved by Daniel Bernoulli in 1733, and in solving it, he introduced a Bessel function into mathematics, although of course he worked with an infinite series and did not name the function involved. Today we write Daniel Bernoulli's solution as

$$y(x,t) \; = \; A\cos(\omega t)J_0\left(2\omega\sqrt{x/g}\right).$$

The rope is fixed at the top, $x = L$, and imposing that condition gives us a formula for the oscillation angular frequency ω:

$$y(L,t)=0 \quad \text{requires} \quad J_0\left(2\omega\sqrt{L/g}\right)=0$$

$$\text{or} \; \left(2\omega\sqrt{L/g}\right)= j_p \; \text{where} \, j_p \; \text{is the pth zero of J}_0.$$

Thus the possible frequencies of oscillation are related to the roots or zeros of the Bessel function. Bernoulli calculated very accurate values for the first two roots. Thus Daniel Bernoulli could use his superposition technique to solve for the motion of a hanging chain in terms of its harmonics, and the shapes taken up by the chain are given by Bessel functions. Equation (12.6) may be used with Bessel functions for $\psi_n(x)$.

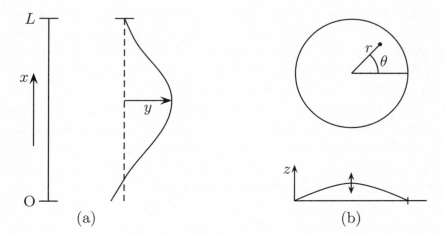

Figure 12.4. (a) The hanging heavy rope problem of Daniel Bernoulli. (b) Vibrating circular membrane, or drum, as considered by Euler. *Figure created by Annabelle Boag.*

Another natural extension of dynamical problems was from the one-dimensional vibrating string to the two-dimensional vibrating membrane, or drum. Solutions for this problem were obtained by Leonard Euler in 1764. For a circular membrane and using polar coordinates r and θ as in figure 12.4, Euler found a solution that may be written as

$$z(r,\theta,t) = A\cos(\omega t)\sin(m\theta)J_m(\omega r/c).$$

Now Bessel functions of higher order have been introduced. The displacement will be zero at the rim of the membrane and that will impose a condition which gives the frequencies of the membrane vibrations in terms of the zeros of the Bessel functions. Euler made some very accurate calculations of a few Bessel function zeros. As usual, the superposition theory may be used to cover all possible vibrations of a drum.

It soon became clear that any problem involving vibrations or waves in systems with circular or cylindrical symmetry would involve Bessel functions. Fourier in 1822 and Poisson in 1823 found Bessel functions appearing in heat-conduction problems with those same symmetries. We already saw in chapter 9 that Airy's calculation of the diffraction pattern formed by light waves leaving a circular aperture involved the J_1 Bessel function. Personally, I spent a long time playing with Bessel functions while working on the way light propagates along optical fibers. (See the book by McLachlan for the viewpoint in 1934.)

Bessel functions have a habit of popping up in all sorts of places. An example is given by a problem of obvious interest to Bessel himself. When describing the motion of a planet in its elliptical orbit, it is necessary to use the eccentric anomaly E and the mean anomaly M. For an orbit with eccentricity ε the two are linked through Kepler's equation:

$$M = E - \varepsilon \sin(E).$$

In the planetary problem, M is given and it is required to find E, something that cannot be done exactly. (See Pask chapter 12 for an introduction to this problem and its analysis by Newton and others.) In 1770, Lagrange showed that a solution to Kepler's equation could be expressed as an infinite series involving Bessel functions:

$$E = M + \sum_{n=1}^{\infty} \left(\frac{2J_n(n\varepsilon)}{n} \right) \sin(nM).$$

This solution has been much studied, including by Bessel himself (see Watson and Dutka).

The message is: if you want to use mathematics in science, be prepared to confront Bessel functions!

12.2.3 Tables

Sometimes an analytical result can be used to make further progress in science, but very often it is numerical results that are needed. (The books by Watson and McLachlan give an incredible assortment of analytical results involving Bessel functions—sums, products, derivatives and differential equations, recurrence relations, integrals and special forms, and limiting cases.) In chapter 4, we saw Lord Kelvin calculating details of terrestrial phenomena, and he firmly and explicitly set out his views in this famous statement:

> In the physical sciences a first essential step in the direction of learning any subject is to find principles of numerical reckoning and practical methods for measuring some quality connected with it. I often say that when you can measure what you are speaking about and express it in numbers you know something about it; but when you cannot measure it, when you cannot express it in numbers, your knowledge is of a meager and unsatisfactory kind.[4]

Thinking of this sort led to the development of mathematical tables. The obvious idea is to tabulate results that may be simply read off rather than recalculated every time they were needed. For many readers, growing up in the computer age, the idea of using tables may seem curious, but as labor-saving devices, they were essential. There were even the common "ready reckoners" which allowed shop assistants and business people to quickly work out quantities required and prices for various goods—a world away from the automatic tills at shop checkouts today.

We have already seen tables constructed by the Babylonians (chapter 2), tables in the great astronomical works by Ptolemy and Kepler (chapter 5), tables of logarithms (chapter 3), and population and annuity tables (chapter 8). (The book edited by Horsburgh takes 15 pages to list tables available in 1915; a more recent history is given by Campbell-Kelly and collaborators.)

You would expect early tables to cover things like logarithms and trigonometric functions, but there is also a wealth of information about numbers themselves, some of it quite strange to our modern eyes. For example, there are tables giving one-quarter of the squares of numbers. So if the number is six, say, the table gives $\frac{1}{4}(6^2) = 9$. In 1887, Joseph Blater published a quarter-squares table for all numbers from 1 to 200,000. (See the articles by McFarland Roegel if you would like to read more about that.) Before you dismiss that as totally weird, let me remind you that we are talking about an era where calculations were done largely without the assistance of a machine. All sorts of clever techniques were invented for making arithmetic easier; one of those techniques involved the use of quarter-square tables. The identity

$$xy = \tfrac{1}{4}(x+y)^2 - \tfrac{1}{4}(x-y)^2$$

tells us how to work out a product of two numbers x and y using the quarter-squares table. Suppose you want 234 times 568 (a tedious task!); we form 234+568 = 802 and 568–234 = 334, and then consult the quarter-squares table to find that 234 times 568 is 160,801–27,889 = 132,912. It takes one simple addition and two easy subtractions along with two table look-ups to get the answer. The same simple set of operations would deal with 119,234 times 78,999 if you had Blater's table on hand. A great amount of ingenuity and mathematical insight was required for the efficient production of mathematical tables, not to mention methods for checking them for errors.

Tables were constructed at a time when the word *computer* meant a person who did calculations. Teams of these "computers" were organized for big projects, and there was a structure of chiefs and supervisors who set out what was to be calculated, decided which techniques to use, and checked the methods to be applied. (The fascinating history of this work is described in articles in the Campbell-Kelly book and in Grier's *When Computers Were Human*.)

Gradually, tables of Bessel functions began to appear, often as part of a larger publication. Bessel himself included tables of $J_0(x)$ and $J_1(x)$ in his memoir on planetary perturbations. Airy gave tables of Bessel functions in his papers, including $J_1(x)$ (actually $2J_1(x)/x$) in his optics paper referred to in section 9.3.1). Lord Rayleigh included a table of $J_0(x)$ and $J_1(x)$ values for x between zero and 13.4 in his 1877 *Theory of Sound*. One of the first extensive tables was published by Ernst Meissel in 1888; he gave $J_0(x)$ and $J_1(x)$ values to an accuracy of twelve decimal places for values of x ranging from 0 to 15.5 in steps of 0.01. A table with an accuracy of twenty decimal places was published in the *Proceedings of the Royal Society* (1900) by William Aldis for x increasing in steps of 0.1 from 0 to 6. Just imagine the work required to produce such tables! (For details of other early calculations, see Watson chapter 20.)

Systematic table preparation and publication was given a boost in 1871 when the British Association for the Advancement of Science established its Mathematical Tables Committee. It will be no surprise to learn that Lord Kelvin was involved. From 1889 onward, a variety of Bessel-function tables were produced, although the type of functions tabulated tended to depend on the particular membership of the committee. (The detailed history of the BA Tables Committee is described in the articles by Mary Croaken.) Eventually, the Bessel Function Subcommittee was formed, and the wonderful Bessel-function tables were published in the British Association series of volumes in 1937, 1952, 1960, and 1964.

For many scientists (me included), the "bible" in the area was "Abramowitz and Stegun." This is the *Handbook of Mathematical Functions*, edited by Milton Abramowitz and Irene Stegun and published in 1964. Digging out my copy, I immediately noticed how worn are the pages of F. W. J. Olver's section on Bessel functions. Extensive tables are given including those for the values of $J_0(x)$ (to fifteen decimal places) and $J_1(x)$ (to ten decimal places), and for their first twenty zeros (to ten decimal places for $J_0(x)$. Much of this material was taken from British Association publications.

Naturally, the arrival of electronic computers meant that it was far simpler to produce mathematical tables. Soon of course they also made such tables redundant; computers can now easily produce values of Bessel functions and other special functions on demand. It might be hard to appreciate the role played by Bessel functions and their tables in the development of science, but it was crucial and involved a level of expertise and dedication that today we can only marvel at. **Calculation 48, tabulating Bessel functions** must find a place in my list of important calculations.

12.3 ABOUT LINEARITY AND BEYOND

The previous two topics are intimately connected with the concept of linearity, which is vitally important when describing a great many physical systems. The next two topics relate to what happens when we look at systems taking us beyond linearity, so a little about the concept of linearity and its implications might be in order. (Those readers familiar with the details of linear systems should skip on to section 12.4.)

Think about a system which has an input p and gives an output q. If the system is linear, we know that

1. if the input is multiplied by some number α then the output will be multiplied by α;
2. if an input p_1 and gives an output q_1 and an input p_2 and gives an output q_2, then an input $(p_1 + p_2)$ and gives an output $(q_1 + q_2)$.

We can write this symbolically by using L to represent the linear system, and then

$$L(p) = q \quad \text{means that } L(\alpha p) = \alpha q \text{ and } L(p_1 + p_2) = q_1 + q_2.$$

For example, if the system is simply "multiply by 3" and the input is numbers, we clearly have a linear system because $3(n + m) = 3n + 3m$ for any numbers n and m. But taking a square root is not a linear process; $\sqrt{9} = 3$ and $\sqrt{16} = 4$ but $\sqrt{(9 + 16)} = \sqrt{25} = 5$ which is not $3 + 4$ as it would be if $\sqrt{}$ was a linear operation.

Very importantly, the calculus operation of differentiation is a linear operation:

$$\text{if } \quad \frac{dp_1(x)}{dx} = q_1(x) \quad \text{and} \quad \frac{dp_2(x)}{dx} = q_2(x) \quad \text{then} \quad \frac{d\{p_1(x) + p_2(x)\}}{dx} = q_1(x) + q_2(x).$$

A similar result holds for higher-order derivatives. Thus we are led to the idea of linear differential equations like equations (12.1) and (12.7). In linear differential equations, each term involves only the unknown function or a derivative; there are no terms involving products like y^2 or $y(dy/dx)$. The linearity property allows us to add solutions (as we have just seen in section 12.1), and this underpins the mathematics discussed in the previous two sections.

The mathematics of linear systems is highly developed, and the linearity properties are enormously helpful in finding general solutions. We saw an example of that in section 12.1. Essentially, we look for a set of basis solutions and then use the linearity property to write any possible solution in terms of them. The only thing required for a particular solution is to find the coefficients or multiplying constants like the a_n in equations (12.2) and (12.3).

Physically, we say that a linear system is defined in terms of its modes, and we will find that the strength or amplitude a_n of a given mode n depends on the particular excitation conditions. (The plucked string is a good example.) There are two properties of such a linear system that should be emphasized.

First, to describe the system in any particular case we need only specify those amplitudes a_n. They will allow us to calculate anything else we want to know. Those amplitudes fully characterize the behavior of the system.

Second, the modes of a linear system behave independently, and the a_n do not change with time; each mode maintains its amplitude or strength and thus contains the same energy at all times. The total energy is the sum of the individual modal energies. For example, a plucked string will have several modes vibrating at the same time (each with its own frequency), and each mode carries a certain amount of energy which remains constant for all time. (Of course, there are energy dissipation effects, but they are not considered here.) Light propagating along an optical fiber is carried in the optical modes of the fiber; each mode carries a fixed amount of energy as it propagates, with that modal energy determined by the light directed into the fiber.

In dynamical systems, particles may be in an equilibrium state and then oscillate around that state if they are displaced. If the forces pulling the particles back toward the equilibrium state depend simply on their displacements, then the system is linear and the equations of motion are linear differential equations. This is the case when Hooke's law holds for elastic systems and springs. The simplest example is a pendulum. If the bob is displaced so the string makes an angle θ with the vertical, the force pulling the bob back toward the $\theta = 0$ position is proportional to θ, and we get the well-known linear equation for the periodic motion of the pendulum.

A large number of physical systems behave in a linear manner, and that is why science has made great progress in many areas. However, there are physical systems that are not linear, and, in fact, if most physical systems are pushed to extremes, they start to lose those beautiful linear properties. (If a pendulum is displaced through very large angles, the exact linearity property is lost.) The question becomes: What happens when a physical system behaves in a nonlinear manner? The exact mathematics of nonlinear systems tends to be very difficult, and the next two sections look at other ways of exploring this area of science.

12.4 A NEW KIND OF EXPERIMENT

Enrico Fermi was one of that rare breed: a first-rate theorist and a superb experimentalist. We met him in section 11.3 in connection with the theory for weak decay processes and in section 11.8, which noted his pioneering experiment on nuclear chain reactions. Fermi had a wide range of interests, and from the very start of his career, he worked on the properties and statistics of systems of many particles. Fermi was intrigued by questions like: Why do we find irreversibility in nature when so many of the fundamental laws are time reversible? Do systems cycle through all possible states (as suggested in the ergodic hypothesis) and tend to equilibrium states with energy equally distributed over all possible degrees of freedom? (That is called the equipartition of energy.) These are difficult questions to answer, and, in 1955, Fermi reported on a new approach for tackling them. It was an approach that led to surprising results that have been the subject of investigation and extension ever since.

Fermi worked with John Pasta and Stanislaw (Stan) Ulam at the Los Alamos laboratories, and their first investigations were published in the now-famous "Studies of Nonlinear Problems." (Perhaps there should have been another author listed—a point I return to in section 12.4.3.)

The particular problem studied in this report has become known as the FPU problem (after the names of the investigators: Fermi, Pasta, and Ulam). The work can best be introduced by quoting from the beginning of the report itself:

This report is intended to be the first in a series dealing with the behavior of certain nonlinear physical systems where the non-linearity is introduced as a perturbation to a primarily linear problem. The behavior of the system is to be studied for times which are long compared to the characteristic periods of the corresponding linear problem.[5]

The linear problem has characteristic oscillations, and the intention was to find out what happens when the linear force terms have nonlinear components added to them. The report immediately notes a difficulty and the way to get around it:

The problems in question do not seem to admit of analytic solutions in closed form, and heuristic work was performed numerically on a fast electronic computing machine (MANIAC I at Los Alamos). [There is a footnote accompanying this sentence which I return to in section 12.4.3.]

Here was the major step: forget trying to find nice analytic solutions to the problem; approximate the equations so that a computer can trace out the solution as a series of numbers. Time as a continuous variable is replaced by a number of discrete intervals. The only way to do this calculation is to use a computer, and some of the first digital computers had been assembled at Los Alamos to support the nuclear bomb program. One of those was the MANIAC—Mathematical Numerical Integrator and Computer. Fermi had access to the MANIAC, and he recognized the possibilities it could bring to science. One of these was to search for an understanding of the fundamental properties of many particle nonlinear systems as the report continues:

The ergodic behavior of such systems was studied with the primary aim of establishing, experimentally, the rate of approach to the equipartition of energy among the various degrees of freedom of the system.

Note that word *experimentally*. Fermi recognized that what they were doing was actually a type of experiment. But instead of watching a real physical system as it moves and changes, they are watching how the output from a set of equations evolves as time goes by in very many discrete steps. Here is a whole new problem for theory: how to suitably approximate the equations to write them in terms of time steps; and then how to choose those steps and make sure that the numbers obtained are a good representation of the exact solution. This is the equivalent of apparatus design for physical experiments, and the literature on numerical analysis and the ideas of computer simulation is now vast.

12.4.1 The FPU Problem: Definition and Results

The FPU problem takes a chain or string of up to 64 particles interacting pairwise through a mixture of linear and nonlinear forces. It is usually thought of in terms of particles joined by springs as in figure 12.5 (a). The particles are all identical, as are the springs. Stretching or compressing a spring by an amount s gives a Hooke's law linear force proportional to s when no nonlinearities are present. (The FPU report talks about a "string" rather than a chain of particles as we do here. The book by Kibble gives a good, simple introduction to the model of a vibrating string in terms of discrete masses.)

Figure 12.5 also shows the case when there are just two particles. In the linear case, there will be two modes for this system, and they are shown in figure 12.5 (b). If the system is set oscillating in one of those forms, it will continue that way for all time. A general displacement of the system will excite a mixture of those two modes, and they will both persist with their energies fixed equal to their initial energies for all time. If the spring forces have a nonlinear component, the linear modes can still be used to describe the system, but now they will be coupled together and can exchange energy. The modes can be thought of like the modes in figure 12.1.

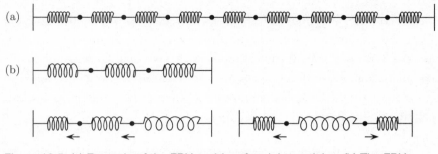

Figure 12.5. (a) Example of the FPU problem for eight particles. (b) The FPU system for two particles and the two modes that are found in the linear system. In the first mode, the particles oscillate together; in the second mode, they oscillate with the same frequency and amplitude but in opposite directions. *Figure created by Annabelle Boag.*

The equations to be solved are Newton's second law of motion (mass times acceleration equals force) for each of the particles. Let the position of the ith particle along the chain be x_i and assume that masses and linear spring constants are both scaled to be 1. Then the FPU report gives the equations to be used for two choices of nonlinearity as

$$\frac{d^2 x_i}{dt^2} = (x_{i+1} + x_{i-1} - 2x_i) + \alpha[(x_{i+1} - x_i)^2 - (x_i - x_{i-1})^2]$$

for a quadratic nonlinearity and

(12.11)

$$\frac{d^2 x_i}{dt^2} = (x_{i+1} + x_{i-1} - 2x_i) + \beta[(x_{i+1} - x_i)^3 - (x_i - x_{i-1})^3]$$

for a cubic nonlinearity.

There is an equation like this for each particle, so for the largest number of particles considered (which was 64) we must put $i = 1, 2, 3, \ldots 64$. Two possible forms for the nonlinear force components are used, and the parameters α and β measure their strengths. Taking α and β as zero gives the linear problem for which the modes can be found and they are given in the FPU report. They are extensions of those given in figure 12.5 (b) but obviously more involved. They may be thought of as much like the modes in figure 12.1, with the first mode having all particles moving together. The calculation is carried out using the linear modes as a basis and discretizing the differential equations so the computer can time-step along the solutions.

The report, "Studies of Nonlinear Problems," now describes the calculation to be made:

> Starting with the string in a simple configuration, for example in the first mode (or in other problems, starting with a combination of a few low modes), the purpose of our computations was to see how, due to nonlinear forces perturbing the periodic linear solution, the string would assume more and more complicated shapes, and, for t tending to infinity, would get into states where all the Fourier modes acquire increasing importance.[6]

So the idea was to use one mode as a starting point and see how the system evolved over time. I can do no better than to quote from the report:

> Let us say here that the results of our computations show features which were, from the beginning, surprising to us. [Even a Nobel Prize winner famous for his intuition was surprised!] Instead of a gradual continuous flow of energy from the first mode to the higher modes, all the problems show an entirely different behavior. Starting in one problem with a quadratic force and a pure sine wave as the initial position of the string [mode one] we indeed observe a gradual increase of energy in the higher modes as predicted (e.g. by Rayleigh in an infinitesimal analysis). Mode 2 starts increasing first, followed by mode 3, and so on. Later on, however, this gradual sharing of energy among successive modes ceases. Instead, it is one or the other mode that predominates. For example, mode 2 decides, as it were, to increase rather rapidly at the cost of all the other modes and becomes predominant. At one time, it has more energy than all the others put together! Then mode 3 undertakes this role. It is only the first few modes that exchange energy among themselves and they do this in a rather regular fashion. Finally, at a later time mode 1 comes back to within one percent of its initial value so that the system seems to be almost periodic. All our problems have at least this one feature in common. . . . It is, there-

fore, very hard to observe the rate of "thermalization" or mixing in our problem, and this was the initial purpose of the calculation.

An example of FPU results for 32 particles is shown in figure 12.6. The system starts in mode 1 and after about 29,000 cycles (measured by the linear mode frequency) it has pretty much returned to that modal state. No wonder they were surprised! Incidentally, as part of this new approach or "mathematical experiment" there is a move toward visual presentation of results and now elaborate graphics is a feature of many papers. A whole set of FPU calculations with varying particle numbers and force parameters show similar results. In terms of the aim of exhibiting the way in which non-linearity leads to mixing of modes and certain statistical properties for the system, the numerical experiment was a failure and the authors write that:

> Certainly, there seems to be very little, if any, tendency towards equipartition of energy among all degrees of freedom at a given time. In other words, the systems certainly do not show mixing.

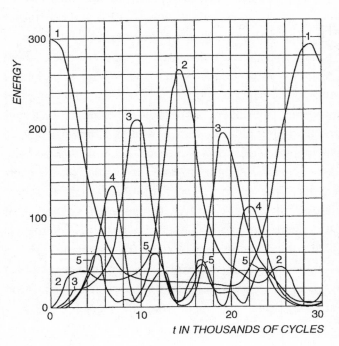

Figure 12.6. Modal energies for the first five modes in an FPU calculation using thirty-two particles and quadratic nonlinear forces. The curves are labeled by the mode numbers. *From E. Fermi, J. Pasta and S. Ulam, "Studies of Nonlinear Problems,"* Los Alamos Document LA-1940 (May 1955).

While the FPU results themselves were surprising, it will come as no surprise to learn that they have led to an enormous amount of work on many diverse aspects of the problem and the system under investigation. The computer power available to modern researchers far exceeds that available for the FPU calculation and thus more extensive results have been generated. Exploring what happens when the various parameters are given different and sometimes extreme values has shown that the oscillations can take on a variety of behaviors including that chaotic form to be discussed in the next section.

The paper by Pettini and colleagues (see bibliography) explains how, as the energy of the initial state is increased, there is a transition first from a regular type of motion to a partially chaotic regime, and then to a fully chaotic situation where full mode mixing is indeed present. The continuum limit (where the number of particles approaches infinity and the space between them becomes zero) leads to a whole new area of science as briefly outlined below in section 12.4.2. This was a calculation that introduced a new approach for studying physical systems, produced intriguing results, and spawned a whole area of science; **calculation 49, FPU and oscillation surprises** surely must find a place in any list of important calculations.

(Any literature search will reveal a great many papers and books about the FPU calculations, their impact, and the new research areas they have generated. I recommend the paper by Porter, Zabusky, Hu, and Campbell for a well-illustrated general introduction. The topic of "computational synergetics" is authoritatively introduced and reviewed by Norman Zabusky, who has played a major part in this area of science. The *Status Report* book edited by G. Gallavotti contains eight up-to-date reviews and also reprints the original FPU report. The review, "The Fermi-Pasta-Ulam Problem: Fifty Years of Progress," by Berman and Izrailev appears in volume 50 of the journal *CHAOS*, which contains a whole series of reviews and progress reports marking their fiftieth anniversary. A paper by Akhmediev shows how the FPU problem is relevant to observations of light waves in optical fibers. Kevrekidis gives an extensive survey of the theory of nonlinear lattice waves, an area of research spawned by the FPU work.)

12.4.2 Continuous Systems

Historically, the initial approach to the stretched-string problem considered a finite number of masses joined by elastic elements (see Kibble for a simple introduction). The limit of an infinite number of masses leads to the continuous stretched-string model as described by the differential equation (12.1). Thus it was natural to ask about the continuous form of the FPU problem. For the quadratic nonlinearity case (non-zero α in equation (12.11)), and with suitably defined displacement u, time variable, and constant parameter δ, the relevant equation is

$$\frac{\partial u}{\partial t} + u\frac{\partial u}{\partial x} + \delta^2 \frac{\partial^3 u}{\partial x^3} = 0.$$

(See the Zabusky paper for further details.) This is known as the KdV equation since it was first written down by D. J. Korteweg and G. de Vries in 1895. They were investigating the propagation of shallow water waves, in which case u refers to height of water in the wave.

The KdV equation has a remarkable solution comprising a wave of given height and shape which propagates unchanged. (This contrasts with the usual linear waves built from many Fourier components and so dispersing since those components travel with different speeds and hence get out of step. In the nonlinear case, the linear dispersion is cancelled out by the nonlinear effects.) These are now termed solitary waves, or solitons. There is a wonderful story of John Scott Russell who, in 1834, followed such a wave on horseback as it traveled along the Union Canal near Edinburgh. The paper by Porter, Zabusky, Hu, and Campbell has a wonderful photograph of surfers on the famous solitary waves traveling along the Severn River in England.

In the FPU calculations, the emphasis was on modal energies rather than actual total displacement shapes for the system (only one of the nine graphs deals with the spatial effects). Thus the surprising recurrence phenomena detected by Fermi, Pasta, and Ulam was not linked by them to waves like solitons, which provide a natural explanation for such phenomena.

The area of soliton physics has grown enormously with many wonderful results and explanations of physical effects driven by nonlinearities. In fact, some of this work might well have made it on to a list of important calculations.

12.4.3 A Mysterious Lady

When the use of the MANIAC computer is mentioned in the FPU report, there is a footnote which reads:

> We thank Miss Mary Tsingou for efficient coding of the problems and for running the computations on the Los Alamos MANIAC machine.[7]

You might think, well that is nice to give a thank-you to an assistant; but was that really enough? Remember, this work was done in the 1950s when electronic computers were in their infancy. To convert a mathematical problem into an algorithm suitable for a machine to follow, then code that algorithm to implement it, and finally to run it on a computer was a major task. Problems that would have been considered immensely difficult to program in 1950 are today tackled by writing a few lines of program in MAPLE or some other computing package, so it may be hard to appreciate just what the early pioneers of computer experiments achieved. (Figure 2 in Thierry Dauxois's article about Mary Tsingou shows the intricate and extensive flow diagram used to program the computer to tackle the FPU problem. Also, according to Dauxois, "Mary Tsingou and John Pasta were the first ones to create graphics on the computer, when they considered the problem with an explosion and visualized it on an oscilloscope."[8]) We saw above that a graphical presentation was the most suitable way to present the FPU results. Clearly, Mary Tsingou was more than just some everyday programmer; it seems likely that she was an essential and integral part of the team leading to the FPU report. Was this another case of a woman in science missing out on some well deserved credit?

Mary Tsingou obtained a bachelor of science degree in 1951 and a master of science in mathematics in 1955. In that era of the Korean War there was a shortage of young men for positions at Los Alamos, and so Mary was part of a group of young women hired to do hand calculations. Obviously she gained entry to the new world of electronic computers and demonstrated a talent for programming. In 1958, she married and became Mary Tingsou Menzel, and she later published scientific papers under that name. More details of this story are in Dauxois's article, and like him, I wonder if really to be fair we should not talk about the FPUT problem.

12.5 DISCOVERING CHAOS

In the preface I wrote that the calculations discussed in this book "are mostly relatively simple (at least in principle, if not always in practice) but still of major significance." In chapter 1 I gave Parson Malthus's calculation of population growth as an example. The calculation itself is almost trivial, but the message contained in the output is profound. It is now time to come full circle and see how time has added to Malthus's work. The calculation I want to tell you about is almost as simple as the original one, and like its predecessor, it has had profound consequences, leading to what might well be called a revolution in the way we think about science and its mathematical description. Because the calculation itself is so important, I will deal with it in detail, and luckily, I believe, in this case everyone can follow those details.

12.5.1 Revisiting Malthus

In chapter 1 I introduced the idea of time periods and population values at the end of each period. Thus p_n is the population at the end of time period n; it could also be the population of generation n when the growth of nonoverlapping populations is being considered. Malthus was interested in the human population, but the growth of populations of all sorts of animals and plants is of major interest today when we seem to be putting so much pressure on our environment. The Malthusian growth problem can be expressed in the simple formula that says the population in one time period is just the growth rate constant g times the preceding or breeding population:

$$p_{n+1} = gp_n. \tag{12.12}$$

This equation has a simple solution telling us how the population changes from a give initial value p_0:

$$p_n = g^n p_0 \quad \text{and as } n \text{ increases:}$$
$$p_n \to 0 \text{ if } g < 1 \quad \text{and} \quad p_n \to \infty \text{ if } g > 1. \tag{12.13}$$

When $g < 1$, there is no growth and the population dies out. When $g > 1$, the population increases without limit. (The special case $g = 1$ has $p_n = p_0$ for all time.) But no population can grow forever, and as Malthus pointed out, the resources for supporting it will come under more and more pressure. Eventually lack of resources—which could mean food, hunting territory, breeding sites, and so on—will cause a decline in the growth of the population. The calculation discussed in this section concerns the mathematical modeling of that process and its implications.

The simplest model of growth limitation may be introduced by defining P to be the maximum population that can be supported by the system under consideration. As the population grows closer to P, the pressures (on food and other resources) will be such that the growth rate g will be reduced. As a simple expression of that we make the change:

$$\text{growth rate } g \quad \text{becomes} \quad \lambda\left(1 - \frac{p_n}{P}\right).$$

Now λ is the new growth rate parameter. When the population is small the growth rate is close to λ, but as the population increases and becomes closer to the possible maximum, the effective growth rate decreases. For example, if $p_n = \frac{1}{4}P$, the effective growth rate is $\frac{3}{4}\lambda$; but if the population increases to be $\frac{3}{4}P$, the growth rate falls to be $\frac{1}{4}\lambda$.

With this change to the description of the growth rate, equation (12.12) for the evolution of the population becomes

$$p_{n+1} = \lambda \left(1 - \frac{P_n}{P}\right) p_n \quad \text{or} \quad \frac{P_{n+1}}{P} = \lambda \left(1 - \frac{P_n}{P}\right) \frac{P_n}{P}.$$

The second form of the equation (obtained from the first by dividing both sides by P) tells us about the population as a fraction of the possible maximum population P. If we call that fraction x, the new population evolution equation may be written in the standard form found in the literature:

$$x_{n+1} = \lambda x_n (1 - x_n). \tag{12.14}$$

This equation is so important that it has a name: the logistic equation. It is important because it proves to be mathematically interesting and because it also has applications in many areas of biology, social sciences, physics, and engineering. It may be hard to accept, but that simple equation (12.14) reveals a whole new world of population behaviors and challenges some of the most cherished assumptions at the heart of science itself.

12.5.2 Exploring the Logistic Equation

There is no closed form solution for the logistic equation, equation (12.14), as there was for simple Malthusian growth in equation (12.13). The strategy must be to explore the results generated by the logistic equation numerically. That is the "evangelical plea" made by Robert M. May in his seminal 1976 paper "Simple Mathematical Models with Very Complicated Dynamics." He urged that "people be introduced to . . . [the logistic equation] . . . early in their mathematical education. This equation can be studied phenomenologically by iterating it on a calculator or even by hand."[9]

When I was a young graduate student in Sydney in the 1960s, Bob May was a brilliant and inspiring associate professor working on plasma physics. He subsequently changed to biology and has become a world-renowned figure in the study of ecology. He has held professorships at Princeton and in England, became the president of the Royal Society, was chief scientific adviser to the UK government, and is now Baron May of

Oxford, where he is a professor. Lord Robert May's 1976 paper is a wonderful exposition of the topic of this section written by one of the founders of the theory.

Suppose we begin by taking the growth constant λ in equation (12.14) to be 2, so that for small populations, the growth rate is almost Malthus's population doubling. For example, for the case of an initial population one-tenth of the possible maximum, we set $x_0 = 0.1$ and equation (12.14) gives $x_1 = 0.18$; the population has almost doubled. If we continue to calculate x_2, x_3, and so on, we discover that the population increases as 0.2952, 0.4161, 0.4859, 0.4996 . . ., but it does not reach the maximum possible value of 1. In fact, the population eventually settles down to the value $x =$ ½. Try substituting $x_n = ½$ in equation (12.14) with $\lambda = 2$ and you will find it returns $x_{n+1} = ½$; the population stays the same.

If we start with a different value for x_0, the population still finishes up being ½ and then stays at that value. So it is easy to see from the equation that a population of 50 percent of the maximum is what is called a steady-state solution. We can interpret this solution as a balance of population and resources. Try a larger population, $x_n = 0.6$ say, and you will find that the population is pulled down by resources pressures to $x_{n+1} = 0.48$. Everything balances for $x_n = 0.5$. I will call this the equilibrium population x_{eq}.

Decreasing the growth constant λ from 2 to 1.5 reduces the equilibrium population to $x_{eq} = ⅓$ while increasing λ to 2.5 increases the equilibrium population to $x_{eq} = 0.6$. In fact, putting $x_{n+1} = x_n = x_{eq}$ into equation (12.14) gives a simple formula for the equilibrium population: $x_{eq} = (1 - 1/\lambda)$.

This is all very satisfying. The logistic equation does indeed model the effect of resource pressures on population growth and tells us that the population and resources are in balance for an equilibrium population which has a simple dependence on the growth rate constant λ. This seems like a good example of mathematical modeling producing numbers that make sense to us for a given physical situation.

12.5.3 A Surprise

Many animal and plant populations "explode," so we might expect some cases to be modeled by a growth constant λ well beyond 2. We know that there is a general equilibrium value $x_{eq} = (1 - 1/\lambda)$, and we have seen that once the population gets to that value it remains there. However, there is a surprise in store for when we move to $\lambda > 3$: if the population deviates even just a little from x_{eq} it now oscillates around that value while moving further and further away from it. The steady state is now an unstable state. What happens next is quite strange.

Bob May said we should experiment, so here we go. Suppose we take $\lambda = 3.2$ for which $x_{eq} = 0.6875$. As an experiment, take an initial population $x_0 = 0.6$. Using equation (12.14), we generate the following sequence of population values (rounded to four decimal places):

0.6, 0.768, 0.5701, 0.7842, 0.5415, 0.7945, 0.5225, 0.7984, 0.5151, 0.7993, . . .

Instead of converging to the steady-state $x_{eq} = 0.6875$, the population is becoming focused on two populations:

$$x_{eq1} = 0.7994555 \text{ and } x_{eq2} = 0.5130445.$$

But now the population does not tend to a single value; it jumps between these two special values. If we had started with a different initial population x_0, the result would still be the same; there are now these two special population values, and the evolving population eventually settles down to oscillate between them.

For $\lambda < 3$, the population settles down to have the same value at every step, and we may say that it has a period of one. For $\lambda > 3$ we find that the population repeats every second step, and this is called period-two behavior. The system is said to have "period doubled."

Population growth and resource pressures are no longer balanced for these larger growth rates, and there is now a boom-and-bust cycle. The population behavior is a little more elaborate than it was but it still makes sense in terms of overshooting and undershooting.

(A technical aside: imposing the condition that equation (12.14) gives the same result every second time leads to a formula for the two special populations. Let $p = \frac{1}{2}(1 + 1/\lambda)$ then x_{eq1} and x_{eq2} are given by $p \pm \sqrt{p(2 - 3p)}$. Putting $\lambda = 3.2$ will give you the two numbers quoted above.)

12.5.4 Even More Surprises

Increasing the growth constant λ just beyond 3.5 produces another surprise. Now it is the two special population values that lose their stability, and instead we find there are four new special values, and the population cycles through them. Thus it repeats every four steps, and we have a period-four situation. The system is said to have period doubled again. The population booms and busts are now getting really complex!

Perhaps you will guess that increasing λ a little more leads to another period doubling and period-eight behavior. Further period doublings occur as λ increases. It is a bizarre story.

There is one final twist; increasing λ beyond around 3.57 leads to a situation where there are effectively an infinite number of periods, and the output from equation (12.14) leaps around with no discernible pattern at all. This has become known as chaos. It is important to note two points. First, putting an x_0 into equation (12.14) will allow us to steadily generate a string of numbers. If anyone repeats that calculation with exactly the same x_0, they will get exactly the same string of numbers. It might be said that these are predictable, or determined, outcomes, but to all intents and purposes they look as though they were generated by some sort of random process.

Second, if the input x_0 is changed only a tiny amount, the generated sequence of numbers will change and soon become nothing like the original sequence. This is called sensitivity to initial conditions. The simple logistic equation has taken us into a weird world.

12.5.5 Discussion

For a long time it was believed that the world obeyed deterministic laws; the pendulum continues to swing in a regular way, and Newton's mathematical description of the solar system led people to think of a clock-work universe. Of course, there were random effects (wind gusts might disturb a projectile on its regular flight path), but they could be added on, and "noise" was to be expected in any real physical system. There had to be averaging to deal with the enormous number of particles in gases and solids, but still, the deterministic equations provided the starting basis for all of that. It had been a shock to find that at the very smallest levels there was a need to use quantum mechanics and its accompanying probability interpretations. Now came the shock that even the simplest deterministic equation related to macroscopic phenomena might produce a chaotic output.

At this point I need to deal with an obvious question: Is all this talk of chaos just based on that one simple equation? The logistic equation is easy to use and allows its output to be understood in mathematical terms (see May's review). It is the simplest nonlinear difference equation. However, it is now established that there is nothing singular or special about the logistic equation, and period doubling and chaotic behavior have been identified in a multitude of cases. Furthermore, the same sort of behavior has been found for continuously changing systems (see the review by Motter and Campbell for an introduction to the early work of Edward Lorenz and others). The theory has been found to apply in real physical situations.

At this point I should offer a few suggestions for further reading (see bibliography for more information). Collections of articles are given in the books edited by Hall (very readable and as published in *New Scientist*) and Cvitanovic (seminal papers and reviews—including the work of Mitchell Feigenbaum showing the generality of the chaos picture and how order may be found in the route to chaos). Good popular books are by Gleick (giving the story of chaos research and the people involved) and Stewart (with a smooth introduction to the relevant mathematical ideas). The book by Moon is a good example of the impact of the discovery of chaos on applied scientists and engineers. The review, "Chaos at Fifty," by Motter and Campbell is a fine introduction to chaos and its discovery in continuous systems. Chapter 11 in the book by Barger and Olsson is a good introduction to nonlinear mechanics and chaos.

The celebrated French scientist and mathematician Pierre-Simon Laplace (1749–1827) was a champion of Newton's methods and showed how refining the use of his gravitational theory could account for many intricate, tiny, and subtle effects observed in planetary motions. His expression of the power of deterministic systems has become famous:

> Given for one instant a mind that could comprehend all the forces by which nature is animated and the respective situation of the beings that comprise it—a mind sufficiently vast to submit these data to analysis—it would embrace in the same formula the movements of the greatest bodies of the universe and those of the lightest atom; for it, nothing would be uncertain and the future, as the past, would be present to its eyes.[10]

We now know that chaos makes a mockery of Laplace's claim, and signs of chaos are beginning to be widely detected. Few could doubt the impact of **calculation 50, simply chaos**.

The literature on and around chaos is now voluminous and diverse. Chaos theory features in Tom Stoppard's play "Arcadia" and a good conclusion for this section might be to quote from the play's program notes written by none other than Professor Robert May:

> There is a flip side to the chaos coin. Previously, if we saw complicated, irregular or fluctuating behavior—weather patterns, marginal rates of Treasury Bonds, color patterns of animals or shapes of leaves—we assumed the underlying causes were complicated. Now we realize that extraordinarily complex behavior can be generated by the simplest of rules. It seems likely to me that much complexity and apparent irregularity seen in nature, from the development and behavior of individual creatures to the structure of ecosystems, derives from simple—but chaotic—rules. (But, of course, a lot of what we see around us is very complicated because it is intrinsically so.)

I believe all this adds up to one of the real revolutions in the way we think about the world.[11]

12.6 APOLOGIES

I must apologize to those readers who work in the many areas of modern science that are not represented in my list of calculations. This chapter has hinted at the importance of probability theory in science and, for example, the studies of phase transitions using the Ising model and other parts of statistical mechanics are sadly missing. There has also been a great increase in methods of data collection and its statistical analysis, which might have been included. But fifty calculations I said, and so just fifty it remains.

Chapter 13

EVALUATION

*in which I suggest how the calculations may be viewed in
various contexts, explain how I came to choose them,
and pick out ten of them to be labeled "great."*

Here is a list of the fifty calculations described in the preceding chapters. The number after the calculation name gives the section in which it is described. The letters and stars refer to attributes of the calculations that I will explain in section 13.2.

1. Malthus on population growth	1.2	a				e*
2. Mesopotamian Pythagorean triples	2.1		b*	c		
3. Archimedes bounds π	2.2	a	b	c*		
4. Fibonacci's presentation in *Liber Abbaci*	3.1			c		e*
5. production of tables of logarithms	3.2	a*		c*	d	e*
6. Euler solves the Basel problem	3.3		b	c		
7. the prime number theorem	3.4		b	c		
8. Eratosthenes measures the earth	4.1	a				e
9. Kelvin and the age of the earth	4.3	a*	b	c		e
10. seismic rays reveal the earth's interior	4.4	a*	b*	c	d	
11. Galileo describes projectile motion	4.5	a		c		
12. tide predictions	4.6	a		c*		e
13. Ptolemy's *Almagest*	5.1	a*		c*	d	
14. Kepler's astronomical calculations	5.2	a*	b	c	d*	
15. Newton's Moon Test	6.1	a	b			e
16. Newton's determination of planetary masses	6.2	a	b			
17. predicting the return of Halley's Comet	6.3	a	b	c*	d	e
18. the discovery of Neptune	6.4	a	b	c		

19. finding the astronomical unit	6.5	a		c		
20. rotating orbits	6.6	a	b*	c		
21. why the night sky is dark	7.1	a	b			
22. state of the universe	7.2	a	b*		d	
23. Hoyle makes carbon	7.3	a	b*			e
24. galaxy rotation and dark matter	7.4	a*	b			
25. escaping gravity	7.5	a	b	c		e
26. Harvey establishes blood circulation	8.1	a*	b*	c		e
27. Halley values annuities	8.2		b	c		e
28. the Hardy-Weinberg law	8.3	a	b*	c		e
29. the mathematics behind the CT scan	8.4	a		c		e
30. scaling from mice to elephants	8.5	a	b*			e
31. light has a finite speed	9.1	a	b			
32. seeing a rainbow	9.2		b	c		e
33. diffraction and the limit to vision	9.3	a		c		
34. light and electromagnetism	9.4	a*	b*	c		
35. photons exist	9.5	a	b	c		
36. bending light	9.6	a	b			
37. atoms really do exist	10.1	a		c		
38. spectral line patterns	10.2	a	b	c		
39. the new mechanics explain atoms	10.3	a*	b	c		
40. discovering the mass of the neutron	10.4	a	c			
41. the electron magnetic moment	11.1		b	c	d*	
42. why we must have a neutrino	11.2	a	b	c		e
43. decays and time dilation	11.3	a		c		
44. quarks tell us particle masses	11.4	a	b		d*	
45. why the sun shines	11.6	a	b			e
46. planning for a bomb	11.7		b			e*
47. strings and Fourier's mathematical poem	12.1	a		c*	d	
48. tabulating Bessel functions	12.2			c	d	
49. FPU and oscillation surprises	12.4	a*	b	c	d	
50. simply chaos	12.5	a*	b*	c*		e

13.1 LEARNING FROM THE CALCULATIONS

The calculations are part of the story of science and mathematics and provide a selection of some of the key advances in these areas. It is possible to take these calculations as a little database that may be used in different ways to extract information about certain topics. I will briefly introduce three examples. In the following discussions, I refer to the relevant calculations by giving their number, for example [12] indicates **calculation 12, tide predictions**.

13.1.1 The Progress of Science

The calculations described above give us a historical tour through physics.

The early calculations show how astronomers like Ptolemy, Copernicus, and Kepler [13, 14] used data to fit a mathematical model of the solar system in order to generate tables for observers to use. Kepler [14] went further, and from his calculations he deduced Kepler's laws, which summarize the dynamical behavior of the planets. Boyle did a similar thing for gases, giving us Boyle's law linking the pressure and volume of a gas.

A new era began when Newton presented his theory of classical mechanics, and then calculations [15–20] were required to check that it accurately accounted for the solar system. It was also used in continuum systems and vibrations of strings and membranes were studied [47]. Eventually, mechanics also led to kinetic theory and an explanation of Boyle's law [37]. The concepts of mechanics could also be used to describe and explore properties of the earth, its surface phenomena [8, 11, and 12], its internal constitution [10], and even its age [9].

It has always been a struggle to find suitable theories for light and explain optical phenomena [31–33, 35]. Classical physics was completed by Maxwell adding his electromagnetic theory to Newton's mechanics and then producing an electromagnetic theory of light [34]. The fundamentals of science were extended with the arrival of the theory of relativity [22, 25, 36, and 43] and the validating of the atomic hypothesis [37]. Calculations showed how atoms and light were connected [38, 39].

Gradually, the building blocks of matter, or the fundamental particles, were discovered and their properties established **[35, 40–44]**. Nuclear physics finally explained some astrophysical mysteries **[23, 45]** and, sadly, gave us terrible new weapons of war **[46]**.

Finally, a new way to do science using mathematical experiments became feasible with the advent of the electronic computer. Calculations using computer simulations **[49, 50]** have taken us back to classical mechanics and shown that there were still strange things to be discovered.

Throughout all of the work based on and around these calculations, there is the central idea that the theoretical results must match the experimental data for the work to be considered as real science. Some modern theories, like "string theory," have yet to produce any testable theory-experiment links and so (quite rightly in my opinion) many people do not accept them as genuine parts of science.

13.1.2 Mathematical Methods

The first calculations usually involved a combination of geometrical ideas and simple arithmetic. Eratosthenes's calculation **[8]** of the radius of the earth is a good example; it also emphasizes the point that only a simple calculation may be required to establish a result of great significance. Over a thousand years later, the use of the transit of Venus to measure the size of the solar system **[19]** still relied on that geometry-arithmetic combination, although now the details are rather more intricate. Similarly, the modeling of the solar system by Ptolemy **[13]** and Kepler **[14]** was based on Euclidean geometry.

Galileo used geometry to represent physical quantities when he constructed what today we call the speed-versus-time diagram **[11]**, so that a geometrical area now measured distance traveled. He too combined arithmetical arguments with his geometrical approach to produce a theory of projectile motion and range data.

The major change came in the seventeenth century when Newton and Leibniz introduced the calculus. Gradually, analytical methods became dominant, although it must be remembered that Newton's approach in the *Principia* is still very much geometrical, and his Moon Test calculation [15] is still in the old geometry-plus-arithmetic tradition. The work of people like Pierre Varignon, Lagrange, Euler, and Laplace created the analytical form of classical mechanics that has come down to us today. Thus the equations of motion were formulated, and the task was to integrate them in particular cases. The prediction of the return of Halley's Comet [17] is a good example of how these new approaches still required great numerical work. The discovery of Neptune [18] was another impressive example, and, in recent times, there have been wonderful calculations of rocket trajectories and spacecraft orbits [25].

While it is still common practice to plot experimental results on a diagram, a new mathematical task also arose: find a formula that fits this data and reveals the key parameters controlling it. A whole range of mathematical and statistical techniques had to be developed. Examples in this area are Lord Kelvin's work on tides [12], the scaling properties of animals [30], Balmer's discovery of order in the hydrogen spectrum [38], and Hubble's study of the spread of galaxies [22].

The use of mathematics to interpret physical data forced researchers to confront inverse problems, which are often mathematically difficult and require sensitive handling of the input data. Thus the discovery of Neptune [18] required the deviations in the orbit of Uranus to be used to predict the existence of planet Neptune, whereas in the direct problem, we would take Neptune as a known planet and calculate how it affected the orbit of Uranus. Other inverse problems arose in seismology [10] and form the basis of CT scans [29], which have revolutionized parts of modern medical science.

The change from geometrical methods to analytical approaches also allowed scientists to manipulate theories and combine different concepts and known results. Perhaps the best example is the way Maxwell combined electric and magnetic phenomena to produce his equations as the basis for electromagnetism [34]. Combining the conservation laws for energy and momentum led to an analysis of beta decay [42], which ultimately led to the discovery of the neutrino. Einstein was able to manipulate his theory of general relativity to show the way in which light paths must bend when they pass near a massive object [36].

To explore theories and fit them to physical data required great skill and effort. A new approach involved analogue devices. Thus Thomas Young explored wave theory using a ripple tank [33], and Lord Kelvin invented a machine for calculating tides [12]. Lord Kelvin (as Sir William Thomson) and Peter Guthrie Tait included in their *Principles of Mechanics and Dynamics* a whole appendix on "Continuous Calculating Machines."

Gradually, the availability of mechanical calculating machines increased, and much of the arithmetical drudgery was eliminated. (See the book by Michael Williams for the full story.) However, the whole nature of the scientific enterprise was changed with the arrival of the electronic computer. Now it was possible to evaluate formulas, fit formulas to data, and even do algebra, so that enormous calculations—like the calculation of particle masses using quark theories [44]—could be carried out without the hours or even months of work previously required for such tasks. The new approach, known as computer simulations, also entered science [49 and 50].

Looking back over the calculations, it becomes apparent that tables have played several important roles. There were tables that recorded mathematical results for other calculators to use: Mesopotamian tables on clay tablets [2], arithmetical tables in the *Liber Abbaci* [4], tables of logarithms [5], and tables factorizing numbers for people like Gauss to use when studying prime numbers [7]. Tables recording population details were important and used by people like Halley to calculate things like annuity values [27]. Astronomical tables have always been important from ancient times and on to Ptolemy [13] and Kepler [14]. Warfare required ballistic tables, and Galileo showed how to calculate them on the basis of his projectile theory [11]. Tables have been of great practical importance; for example, Lord Kelvin showed how to calculate them for tides [12]. As analytical methods began to dominate theories, it was necessary to produce tables of special functions, and those for Bessel functions are an outstanding example [48]. It is easy in this age of the electronic computer to forget the parts played by tables in almost every area of science. (The history of table-making is comprehensively covered in the book edited by Campbell-Kelly and colleagues.)

In this book, I have been concerned with the interplay of theory and experiment, but the interplay of pure and applied mathematics has also played an important part in the advancement of science. Much of the mathematics used by scientists has been borrowed from pure mathematics, and in turn, science has thrown up new mathematical challenges and inspired mathematicians to develop whole new areas of interest.

13.1.3 People

It is no surprise to find the giants of science appearing as some of my chosen calculators: Archimedes [3], Newton [15, 16, 17, 20, and 32], Gauss [7], Maxwell [34], Einstein [20, 35, 36, and 43], and Bohr [39]. More of their calculations might have been included, and Archimedes, Gauss, and Maxwell in particular are underrepresented. Looking back, it does surprise me that Euler, Fourier, Laplace, Planck, and John von Neumann are not more prominent.

The calculations reveal what a remarkable man Edmond Halley was. Many people are aware of the fact that Halley guided the publication of Isaac Newton's *Principia*, and, of course, he is famous for his work on comets [17]. He also appears in the calculation of the astronomical unit [19], in early actuarial work [27], and in discussion of the dark night-time sky problem [21]. Halley was born in 1656 and died in 1743, and so he unfortunately became one of the people who were overshadowed by Newton (Robert Hooke was another). Halley was mathematician enough to be appointed Savilian Professor of Geometry at Oxford; he was an astronomer of note and first detected stellar motions in 1718 (incidentally using Ptolemy's tables); he became Astronomer Royal; he had interests in subjects as diverse as archaeology and biology; and he was practical enough to be commissioned as a naval captain and command a survey ship from 1698 to 1700. (See Ronan for further details.)

The calculations also reveal how important Sir William Thomson, later Lord Kelvin, was for nineteenth century science. He too was just a little overshadowed—this time by James Clerk Maxwell. Kelvin is the central figure for calculations [9] and [12], but his influence is apparent in other cases, like **[21]** and **[45]**. He was one of the great classical physicists working across fields such as electromagnetism and thermodynamics. It was also Kelvin who championed the importance of numerical studies and showed how analogue devices could be constructed to carry out difficult and extensive calculations. His textbook written with Peter Guthrie Tait was highly influential for many years.

For me, the calculations also identify other people who deserve to be better known for their work, which might not be seen as glamorous as some other cases, but which provided vital, key advances. Mathematically, I can point to Napier and Briggs for their work on logarithms, and to Daniel Bernoulli for founding the methods of linear mathematics **[47]**. Harvey's work on blood circulation **[26]** changed medicine forever. The Reverend John Michell conceived the experiment using attracting spheres to measure gravitational forces (now associated with Henry Cavendish since he first carried out the experiment). Michell speculated about the effects of gravitation on light leading to the possibility of dark stars (and now black holes) **[25]** and the bending of light near massive bodies **[36]**.

Today, many magazines and television programs tell us about the remote reaches of the universe (and often give wild speculations about its nature and origins), but how many people know the names Oldham, Cormack, and Hounsfield? They too tell us about regions that are difficult, if not impossible, to access, but their work is also of paramount practical importance for our actual survival. Richard Oldham made giant strides in seismology **[10]** so that finally we could know what is inside our Earth. Similarly, the work of Allan Cormack and Godfrey Hounsfield led to the CT scan **[29]** and a revolution in medical science allowing us to see inside the body without dissecting it.

Looking over the calculations shows that it was not always the famous scientist working in the great centers of research who made remarkable discoveries. Andrija Mohorovičić (1857–1936), in a 1910 paper, described the earth's crust-mantle discontinuity **[10]**, which is still of major interest to earth scientists and exploration engineers. His name lives on as this discontinuity is known as the "Moho," although it would be interesting to know how many people actually understand the origin of this term. Who would have thought that someone (Max Kleiber, [30]) working in a department of animal husbandry would find a place in a book like this? Similarly, many people might know the term *Balmer series* for some of the spectral lines of the hydrogen atom [38], but not know that it comes from the name of an obscure Swiss schoolmaster, Johann Jakob Balmer (1825–1898), who brilliantly identified a pattern in those spectral lines. Anyone studying genetics will come across the Hardy-Weinberg law [28]; however, probably few of them will realize that, strangely, the Hardy referred to was not a biologist but actually the wonderfully interesting pure mathematician G. H. Hardy.

Surveying the calculations, I find that sadly there are only three women playing central roles, perhaps because I have set and struggled with the fifty calculations limit. Inge Lehmann (1888–1993) made the calculations that led to our picture of the earth's interior core having a solid center with a liquid surround [10]. Vera Rubin was a pioneer for women in astronomy, and her work on matter rotations in galaxies [24] led to one of the great puzzles of modern science: What is dark matter? The experiences of both of these women show how difficult it was for them to gain acceptance in the male-dominated scientific-research world.

The third woman is Lise Meitner (1878–1968), whose seminal work on nuclear fission [46] played a part in the progress toward nuclear weapons (which she regretted). Meitner was born in Vienna, Austria, and became the first woman to receive a doctorate in physics from the University of Vienna. She studied and worked in Germany until 1938 when she fled to Sweden to escape the Nazi regime. Lise Meitner was a remarkable woman who made a career in nuclear physics at a time when women were expected to keep to "traditional roles" as this statement by Max Planck illustrates: "Generally it cannot be emphasized enough that nature herself prescribed to the woman her function as mother and housewife, and the laws of nature cannot be ignored under any circumstances without grave danger."[1] Meitner never married. As a physicist she earned the support, respect, and friendship of the great scientists of her era: Bohr, Boltzmann, Einstein, Pauli, and Planck. (Despite the above quote, Planck employed Meitner and became personal friends with her; in turn, according to biographer Patricia Rife, Meitner greatly respected and admired Planck and enjoyed his company.) Lise Meitner worked extensively with Otto Hahn, her expertise in physics complementing his abilities as a chemist. In later years, Hahn did everything possible to downplay her influence and push her into the background. It is a mystery and a scandal that Otto Hahn received the Nobel Prize for chemistry in 1944; Lise Meitner was never given that honor despite being nominated by many people, including Niels Bohr in 1946, 1947, and 1948. She did receive many other awards, held senior positions, and element 109 is named Meitnerium in her honor. If you wish to know about the difficulties and injustices faced by pioneering women scientists, read the story of Lise Meitner—but I warn you, it may make you feel angry, depressed, astonished, and maybe, as a man, a little ashamed. (Much has been written about Lise Meitner, and there are books by Barron and comprehensive biographies by Rife and Sime.)

One gets the feeling that women's contributions were often played down, like the work of Mary Tsingou in the FPU calculations [49]. Perhaps the best known example is the way Rosalind Franklin was treated by Crick and Watson in the scramble to gain the glory for the discovery of the structure of DNA. Even when they are successful, women can be belittled in awful ways; when Dorothy Crowfoot Hodgkin was awarded the 1964 Nobel Prize for her work on the structure of complex biological molecules, it was reported in the *Daily Mail* under the headline "Nobel Prize for British Wife." (By the way, Margaret Thatcher, the British prime minister, was a pupil of Dorothy Crowfoot Hodgkin at Oxford University, although I am not sure how that helped the cause of women in science!) It is hard not to feel a sense of anger and shame when reading such things, and I am sorry that I have not included more women in my list of important calculators.

In those hidden areas where women have made significant calculations, we must include the people making detailed numerical calculations and especially those in the teams of calculators whose work was vital in the two hundred years up to around 1950 when electronic computers became more available. We saw Madame Nicole-Reine Étable de Labrière Lepaute working alongside Clairaut and Lalande through 1758 to calculate the returning path of Halley's Comet [17]—and typically getting little credit for her contribution. (I highly recommend the writings of David Grier for anyone wishing to learn more about this fascinating period of mathematical work and the part played by women mathematicians.) In chapter 12, I mentioned the tables in the famous "Abramowitz and Stegun" *Handbook of Mathematical Functions*. It is not widely known that Milton Abramowitz died in 1958 and it was the female contributor, Irene Stegun (1919–2008) who saw the project through to publication in 1964. Let us hope that women are more widely and fairly represented in any future list of great calculations.

13.2 ASSESSING THE CALCULATIONS

In 1798, Emperor Napoleon Bonaparte led an army to Egypt. After one particular battle, he took the opportunity (as most visitors to Egypt do) to see the Great Pyramid. While some of his officers climbed to the top, the emperor (sensibly?) decided to stay at the bottom. To pass the time, Napoleon calculated that the stone used to construct the pyramid could be used to build a wall 3 meters tall and 0.3 meters wide around all of France. (His result is reasonably correct too.) You probably find the story surprising, but Napoleon did claim to be something of a mathematician and maintained that "the advancement and perfection of mathematics are intimately connected with the prosperity of the State."[2] Certainly it is a curious result, and it emphasizes just how much stone had to be moved to build the pyramid. But is it a significant or important calculation? Why did I not consider it when forming my list of important calculations?

It is now time to reveal how I evaluated my list of calculations; then we will see if I can narrow it down to just ten great calculations. My criteria for including a calculation are based on five attributes that I label (a)–(e). These are my criteria, and I accept that a strong personal preference or bias is involved both in setting these criteria and in applying them.

Criterion (a). The calculation marks a turning point in science, a breakthrough. Usually it opens up a new field for investigation.

Example: Calculations **[35]** revealing that light may be described in terms of photons.

Criterion (b). The calculation reveals something unexpected and surprising. It often brings gasps of admiration!

Examples: Newton's calculation [16] of the masses of planets or the discovery of chaos [50].

Criterion (c). The calculation is mathematically or computationally innovative and shows the way for future work.

Examples: Archimedes's calculation of π [3] or the production of tables of logarithms [5].

Criterion (d). It is a heroic calculation.

Example: Kepler fitting data to an orbit for Mars [14].

Criterion (e). The calculation had a major social or intellectual impact or an influence on national affairs.

Example: Evaluating uranium fission processes for weapons production [46].

Napoleon's calculation does not score very well using those criteria, although (b) might be suggested by some people.

In the list of calculations given above, I have indicated how I have assigned these attributes to them, with a star indicating an extreme case. Now I can go through the list and pick out ten great calculations. Actually, that is not at all easy to do, and at this point I wonder whether I have set myself an impossible task—probably a silly one too, but it is fun to try! Here are my choices.

13.2.1 Great Calculations

3. Archimedes Bounds π

Archimedes showed how a physical constant could be evaluated using mathematics and how an iterative method could be used to generate an answer to any required degree of accuracy. The idea of placing ever-tighter bounds on the value of π and the actual numerical work were brilliantly carried out.

5. Production of Tables of Logarithms

The idea that multiplications could be replaced by additions was not just brilliant, it was one that changed the nature of computational tasks for many centuries. Only later in the twentieth century, when calculating devices became widely available, did the value of logarithms dissipate. The calculations to provide tables of logarithms can truly be described as heroic.

10. Seismic Rays Reveal the Earth's Interior

Until these calculations were made, it would have been assumed that there was no way to find out what was deep inside the earth. Certainly volcanoes had been observed, and the destructive power of earthquakes was well documented. But the turning of knowledge of seismic waves into information about the earth's interior required a major leap in theory and extremely difficult calculations. This work solved an inverse problem and showed the way for a vast new exploration of the structure and properties of planet Earth that continues today.

14. Kepler's Astronomical Calculations

Kepler's dedication and persistence in his work on understanding the solar system inspire the utmost admiration. He was able to deduce his three laws that exhibit the underlying order in planetary motions. It was Kepler who made the monumental step involving the discrediting of the old Greek idea that planetary motions must involve circles, an idea that had dominated astronomy for centuries. If any calculation deserves the label heroic, it is surely Kepler's calculation of the elliptical orbit of Mars. A new standard was set for scientific working and the comparing of theory and observation.

17. Predicting the Return of Halley's Comet

Newton demystified the nature of comets and suggested that they were not harbingers of great events and tragedies, but regular physical bodies whose orbits might be understood like those of the planets. The difficulty is that only a tiny part of a comet's orbit may be observed and it then moves into the outer reaches of the solar system and thus disappears for many years. The prediction of the return date for a comet was a vital test for Newton's universal theory of gravitation. The calculations required for such a prediction must have seemed daunting in the extreme, but Clairaut and his assistants refined Halley's orbit prediction to give one of the most spectacular and public confirmations of the validity of gravitational theory. (An aside: the discovery of the planet Neptune [18] was a close rival for this spot.)

20. Rotating Orbits

Newton needed properties of motion under an inverse square law to compare with observations in order to confirm that gravity is a force of that type. He was able to prove that for an inverse square law force, the planetary orbits are ellipses, while for other force laws, the orbit might be thought of as an ellipse but it would also slowly rotate in space. Using innovative mathematics and perturbation theory, Newton produced a formula for the rotation rate showing how it increased as deviations from inverse square increased. Comparing planetary data, he was able to mount a strong argument for the mathematical form of his theory of gravity. Many astronomers calculated the variations in orbits due to perturbations, and a small error in the calculated orbit of Mercury stubbornly remained. It was a compelling result in Albert Einstein's general theory of relativity that explained the discrepancy in terms of an orbit rotation that may be calculated using Newton's formalism. Brilliant mathematics, brilliant physics, and a precise numerical calculation gave Einstein one of his greatest triumphs.

26. Harvey Establishes Blood Circulation

A simple combination of data about the heart—its rate of beating and its volume—allows us to estimate blood flows and the amount of blood in the human body. The conclusion drawn by William Harvey (although he placed less reliance on this calculation than he did on other evidence) was that blood is not continuously manufactured in amazing quantities in the liver, but that it circulates around the body. This overturned centuries of medical dogma and opened the way for a whole new approach to medicine and an understanding of the physiology of animals. If ever a virtually trivial calculation could be said to have a profound impact, it is surely Harvey's on the pumping of blood by the heart.

34. Light and Electromagnetism

Establishing the nature of light has been one of the great crusades in physics. Describing light as a wave motion is enormously successful, but always leaves us with questions about the nature of the optical waves. The big step was made by Maxwell using his theory which combined, or unified, electric and magnetic phenomena. In the first part of the calculation, Maxwell showed how to manipulate the equations forming that theory to obtain a wave equation, which describes the propagation of electromagnetic waves in space and through media. He was able to use parameters obtained from experiments involving electricity and magnetism to calculate the speed of those waves and thus to show that the speed of electromagnetic waves is the same as the measured speed of light. Few steps in physics could be said to have such a profound effect on both physics and technology as Maxwell's discovery that light is an electromagnetic wave.

39. The New Mechanics Explain Atoms

Modern science is largely based on the atomic hypothesis and its use in explaining the properties of matter and the processes that govern the way our world works. A key step was the explanation of how light and matter interact, and understanding the spectral lines produced by different elements was a central challenge. The result was a new type of mechanics, quantum mechanics, that must be used at the atomic level. The use of the Schrödinger equation to accurately calculate the spectrum for hydrogen represents one of physics' greatest triumphs.

50. Simply Chaos

The last fifty years have seen a revision in our thinking about the output from deterministic systems and the way in which apparent randomness can be present in the physical world. The ideas are now widespread in various types of theories. The calculations used to explore the strange properties of the output generated by the logistic equation is a good example of the power of a relatively simple calculation. The further development and applications of the mathematics and the corresponding experiments have been extensive.

13.3 FINAL WORDS

I began this book by talking about the interplay between theory and observation and experiment. It is easy to have faith in a beautiful theory, as with Einstein and his theory of relativity; but we must always remember that experiment is the final arbiter, as Einstein also recognized when he explained that while "no amount of experimentation can ever prove me right; a single experiment can prove me wrong."[3] On the other hand, Sir Arthur Eddington (who might be called Einstein's disciple) turned things around when he suggested that "it is also a good rule not to put overmuch confidence in the observational results that are put forward until they are confirmed by theory."[4] Eddington might have had a lively debate with Fred Hoyle who said, "I don't see the logic of rejecting data just because they seem incredible."[5] Opinions may vary a little, but in the end, scientific methodology remains beautifully summarized in the words of Robert Millikan (taken from his address on receiving the 1923 Nobel Prize for his work on measuring the fundamental unit of electric charge):

> Science walks forward on two feet, namely theory and experiment. . . .
> Sometimes it is one foot that is put forward first, sometimes the other, but
> continuous progress is only made by the use of both—by theorizing and
> then testing, or by finding new relations in the process of experimenting
> and then bringing the theoretical foot up and pushing it on beyond, and
> so on unending.[6]

I hope you can look back and see the various ways in which calculations and experiments have moved science along in cooperation.

Finally there are my ten choices for the great calculations. It may be a futile exercise, and I am sure every reader will come up with a different set. However, it does make one think about the role played by outstanding calculations in science, and I am pleased if you have been inspired to consider your own experiences. For example, some older readers might, like me, even remember using log tables at school or in college, although we may have forgotten what incredible labor-saving devices they were.

During my lifetime I have been privileged to see a return of Halley's Comet and observe a transit of Venus. On both occasions, I am sure that an appreciation of the superb background calculations enhanced the experience. Maybe we can all smile with pleasure when we read (in section 11.7.1) about Fritz Houtermans's boast to his girlfriend that he knew why the stars shine. Perhaps you too will recall events in your life that relate to some of the calculations I have described. If you have found my discussions of the calculations in some way stimulating, entertaining, and maybe even provocative, then I will feel satisfied.

NOTES

CHAPTER 1: INTRODUCTION

1. Albert Einstein, "The Fundamentals of Theoretical Physics," in *Essays in Physics* (New York: Philosophical Library, 1940).

2. Attributed to Kant, though it is not clear where he used these exact words. Certainly they are a good summary of the extensive discussions in his *The Critique of Pure Reason* (Chicago: Encyclopaedia Brittanica, 1990).

3. "Darwin, C. R. to Fawcett, Henry, 18 Sept [1861]," *Darwin Correspondence Project*, http://www.darwinproject.ac.uk/entry-3257 (accessed April 17, 2013).

4. Albert Einstein in a conversation reported in Werner Heisenberg, *Physics and Beyond: Encounters and Conversations* (London: George Allen & Unwin, 1971), p. 63.

5. Galileo in *The Assayer* (1623). See M. A. Finocchiaro, ed., *The Essential Galileo* (Indianapolis: Hackett Publishing, 2008).

6. For an introduction, see J. Baggott, *Farewell to Reality: How Modern Physics Has Betrayed the Search for Scientific Truth* (New York: Pegasus, 2013).

7. For a discussion and range of examples, see Monwhea Jeng, "A Selected History of Selection Bias in Physics," *American Journal of Physics* 74 (2006): 578–83.

8. Thomas Malthus, *Essay on the Principle of Population as It Affects the Future Improvement of Society* (London: St Paul's, 1798).

9. Charles Babbage in a letter to the poet Tennyson in W. F. Bynum and R. Porter, eds., *Oxford Dictionary of Scientific Quotations* (Oxford: Oxford University Press, 2005), p. 34.

CHAPTER 2: ANCIENT MATHEMATICS

1. Euclid, *The Thirteen Books of the Elements*, trans. Sir Thomas Heath (New York: Dover, 1956).

2. Archimedes in Sir Thomas Heath, *A History of Greek Mathematics*, vol. 2, *From Aristarchus to Diophantus* (Oxford: Clarendon, 1921). See pages 50–56 for Archimedes's "Measurement of the Circle."

3. Ibid.

4. Ibid.

5. Euclid, *The Thirteen Books of the Elements*.

CHAPTER 3: STEPS INTO MODERN MATHEMATICS

1. Keith Devlin, *The Man of Numbers: Fibonacci's Arithmetic Revolution* (London: Bloomsbury, 2011), p. 86.

2. John Napier, *Mirifici Logarithmorum Canonis Descriptio* (1614).

3. Ronald Calinger, *A Contextual History of Mathematics* (Upper Saddle River, NJ: Prentice Hall, 1999), p. 485.

4. Pierre-Simon de Laplace quoted in ibid.

5. Translation from the original Latin provided by Mark Hall.

6. See "Are These the Most Beautiful?" by David Wells in *Mathematical Intelligencer* 12 (1990): 37–41.

7. Pierre Simon de Laplace quoted in Robert A. Nolan, *A Dictionary of Quotations in Mathematics* (London: McFarland, 2002), p. 115.

8. Carl Friedrich Gauss quoted in Nolan, *Dictionary of Quotations in Mathematics*, p. 189.

9. Thomas Henry Huxley, "Scientific Education—Notes on an After-Dinner Speech," *Macmillan's Magazine* (June 1868).

10. Howard Eves, *Great Moments in Mathematics (before 1650)* (Washington, DC: Mathematical Association of America, 1980), p. 116.

CHAPTER 4: OUR WORLD

1. Bishop James Ussher quoted in D. R. Oldroyd, *Thinking about the Earth: A History of Ideas in Geology* (Cambridge, MA: Harvard University Press, 1996), p. 49.

2. Sir William Thomson (Lord Kelvin), "On the Secular Cooling of the Earth," *Transactions of the Royal Society of Edinburgh* 23 (1862).

3. Quoted in Ivan Ruddock, "Lord Kelvin, Science and Religion," *Physics World* (February 2004): 56.

4. Thomas Henry Huxley, *Quarterly Journal of the Geological Society*, vol. 25 (1869), quoted in W. F. Bynum and R. Porter, eds., *Oxford Dictionary of Scientific Quotations* (Oxford: Oxford University Press, 2005), p. 530.

5. Ernest Rutherford (1904), quoted in Bynum and Porter, *Oxford Dictionary of Scientific Quotations.*, p. 530.

6. R. D. Oldham, "On the Propagation of Earthquake Motion to Great Distances," *Philosophical Transactions of the Royal Society of London, Series A* 194 (1900): 135–74.

7. R. D. Oldham, "The Constitution of the Interior of the Earth as Revealed by Earthquakes," *Quarterly Journal of the Geological Society* 62 (1906).

8. Quoted in Martina Kölbl-Ebert, "Inge Lehmann's Paper: 'P' (1936)," *Classic Papers in the History of Geology* 24 (2001): 263.

9. Inge Lehmann, "Seismology in the Days of Old," *Eos Transactions American Geophysical Union* 68 (1987): 33–35.

10. Quoted in Marijan Herak, "Mohorovičić Discontinuity," *Andrija Mohorovičić Memorial Rooms*, http://www.gfz.hr/sobe-en/discontinuity.htm (accessed March 12, 2014).

11. Craig Jarchow and George Thompson, "The Nature of the Mohorovičić Discontinuity," *Annual Review of Earth and Planetary Science* 17 (1989): 475.

12. Galileo Galilei, *The Assayer* (1623). See M. A. Finocchiaro, ed., *The Essential Galileo* (Indianapolis, IN: Hackett, 2008).

13. Galileo Galilei, *Two New Sciences*, trans. Stillman Drake (Madison: University of Wisconsin Press, 1974), p. 190.

14. Ibid., p. 197.

15. Ibid., p. 154.

16. Ibid., p. 77.

17. Ibid., p. 222.

18. Ibid., p. 268.

19. Ibid., p. 245.

20. Sir William Thomson (Lord Kelvin) and P. G. Tait, *Principles of Mechanics and Dynamics*; formerly titled *Treatise on Natural Philosophy* (New York: Dover, 1962). A republication of the last revised edition of 1912.

21. Bruce Parker, "The Tide Predictions for D-Day," *Physics Today* 64 (September 2011): 38.

CHAPTER 5: THE SOLAR SYSTEM: THE FIRST MATHEMATICAL MODELS

1. G. J. Toomer, "Ptolemy," in *Dictionary of Scientific Biography*, ed. C. C. Gillespie (New York: Charles Scribner's Sons, 1975), p. 187.

2. Ptolemy, *Almagest*, trans. G. J. Toomer (London: Duckworth, 1984), p. 57.

3. Isaac Newton quoted in D. T. Whiteside, "Newton's Lunar Theory: From High Hope to Disenchantment," *Vistas in Astronomy* 19 (1976): 317–28.

4. From the introduction to M. Campbell-Kelly, M. Croaken, R. Flood, and E. Robson, eds., *The History of Mathematical Tables* (Oxford: Oxford University Press, 2003).

5. Quoted in C. M. Linton, *From Eudoxus to Einstein: A History of Mathematical Astronomy* (Cambridge: Cambridge University Press, 2004), p. 121.

6. Ibid., p. 126.

7. Quoted in O. Gingerich, "Kepler," in Gillespie, *Dictionary of Scientific Biography*, p. 290.

8. Ibid., p. 295.

9. Ibid., p. 297.

CHAPTER 6: THE SOLAR SYSTEM: INTO THE MODERN ERA

1. Newton quoted in Colin Pask, *Magnificent* Principia*: Exploring Isaac Newton's Masterpiece* (Amherst, NY: Prometheus Books, 2013), p. 395.

2. Peter H. Cadogan, *The Moon: Our Sister Planet* (Cambridge: Cambridge University Press, 1981).

3. Quoted in I. B. Cohen, "Newton's Determination of the Masses and Densities of the Sun, Jupiter, Saturn, and the Earth," *Archive for the History of Exact Sciences* 53 (1998): 83.

4. Daniel Defoe, *A Visitation of the Plague* (London: Penguin Books, 1986), originally published in London in 1722.

5. Isaac Newton, *The Principia*, bk. 3 (Amherst, NY: Prometheus Books, 1995), pp. 396–401.

6. Halley in a letter to Newton quoted in C. B. Waff, "Predicting the Mid-Eighteenth-Century Return of Halley's Comet," in *The General History of*

Astronomy, vol. 2, *Planetary Astronomy from the Renaissance to the Rise of Astrophysics Part B: The Eighteenth and Nineteenth Centuries*, ed. R. Taton and C. Wilson (Cambridge: Cambridge University Press, 1989), p. 70.

7. From an address by Joseph-Jérôme Lefrançois de Lalande quoted in Waff, "Predicting the Mid-Eighteenth-Century Return of Halley's Comet," p. 69.

8. Quoted in N. R. Hanson, "Leverrier: The Zenith and Nadir of Newtonian Mechanics," *Isis* 53 (1962): 361.

9. Quoted in M. Grosser, *The Discovery of Neptune* (New York: Dover, 1979), p. 93.

10. Quoted in A. Van Helden, "Measuring Solar Parallax: The Venus Transits of 1761 and 1769 and Their Nineteenth Century Sequels," in *The General History of Astronomy*, vol. 2, *Planetary Astronomy from the Renaissance to the Rise of Astrophysics Part B: The Eighteenth and Nineteenth Centuries*, ed. R. Taton and C. Wilson (Cambridge: Cambridge University Press, 1989), p. 154.

11. Edmond Halley, "A New Method of Determining the Parallax of the Sun," *Philosophical Transactions* 39 (1716): 454–56. Translation available at http://eclipse.gsfc.nasa.gov/transit/HalleyParallax.html (accessed April 12, 2014).

12. Data from C. M. Linton, *From Eudoxus to Einstein: A History of Mathematical Astronomy* (Cambridge: Cambridge University Press, 2004), p. 49 (see table 12.1).

13. A. Pais, *Subtle Is the Lord: The Science and Life of Albert Einstein* (Oxford: Oxford University Press, 1982), p. 253.

CHAPTER 7: THE UNIVERSE

1. Lucretius, *De Rerum Natura*, written around 50 BCE and translated by C. H. Sisson as *The Poem on Nature* (Manchester, UK: Carcanet, 2006). Reprinted with permission.

2. Digges quoted in E. R. Harrison, *Cosmology: The Science of the Universe*, 2nd ed. (Cambridge: Cambridge University Press, 2000), p. 492.

3. Kepler quoted in Harrison, *Cosmology*, p. 493.

4. Harrison, *Cosmology*, p. 505.

5. E. Hubble, "A Relation between the Distance and Radial Velocity among Extra-Galactic Nebulae," *Proceedings of the National Academy of Sciences* 15 (1929): 168–73.

6. Isaac Newton, "General Scholium," in *The Principia*, bk. 3 (Amherst, NY: Prometheus Books, 1995), p. 440.

7. J. A. Wheeler, *Geons, Black Holes and Quantum Foam* (New York: Norton, 2000).

8. R. J. A. Lambourne, *Relativity, Gravitation and Cosmology* (Cambridge: Cambridge University Press, 2010), p. 276.

9. Taken from a page of Fred Hoyle's notebook shown in *Fred Hoyle: An Online Exhibition*, http://www.joh.cam.ac.uk/library/special_collections/hoyle/ (accessed May 12, 2014).

10. Fred Hoyle, *Astronomy and Cosmology: A Modern Course* (San Francisco: W. H. Freeman, 1975), p. 402.

11. Fred Hoyle, *Home Is Where the Wind Blows: Chapters from a Cosmologist's Life* (Mil Valley, CA: University Science Books, 1994), p. 266.

12. Vera C. Rubin, "The Rotation of Spiral Galaxies," *Science* 220, no. 4604 (1983).

13. J. P. Ostriker, P. J. E. Peebles, and A. Yahil, "The Size and Mass of Galaxies, and the Mass of the Universe," *Astrophysical Journal* 193 (1974): L1–L4.

14. Vera C. Rubin, "Seeing Dark Matter in the Andromeda Galaxy," *Physics Today* (December 2006).

15. John Michell, sec. 16 in "On the Means of Discovering Distance . . . In a Letter to Henry Cavendish," *Philosophical Transactions of the Royal Society of London* 74 (1784): 35–57.

16. Freeman Dyson, "Chandaransekar's Role in 20th-Century Science," *Physics Today* 63 (December 2010): 47.

CHAPTER 8: ABOUT US

1. Quoted in the biographical note preceding William Harvey, *An Anatomical Disquisition on the Motion of the Heart and Blood in Animals, A Translation of the 1628 Original* Exercitatio Anatomica de Motu Cordis et Sanguinis in Animalibus (Chicago: Great Books of the Western World, Encyclopedia Britannica, 1990).

2. Ibid.

3. Harvey, *Anatomical Disquisition*, p. 286.

4. Ibid.

5. Harvey in the dedication to his colleagues in *Anatomical Disquisition*.

6. Andrew Gregory, *Harvey's Heart: The Discovery of Blood Circulation* (Duxford, UK: Icon Books, 2001), p. 118.

7. James R. Newman in his commentary introducing the excerpt from *Natural and Political Observations Mentioned in a following Index and Made upon the Bills of Mortality* (published by John Graunt in London, 1662), in *The World of Mathematics*, vol. 3, ed. James R. Newman (London: George Allen and Unwin, 1990), p. 1416.

8. John Graunt in *Natural and Political Observations Mentioned in a following Index and Made upon the Bills of Mortality* (published by John Graunt in London, 1662), reproduced in *The World of Mathematics*, vol. 3, ed. James R. Newman (London: George Allen and Unwin, 1990), p. 1433.

9. Edmond Halley, "An Estimate of the Degrees of the Mortality of Mankind, Drawn from Curious Tables of the Births and Funerals at the City of Breslaw, with an Attempt to Ascertain the Price of Annuities upon Lives," *Transactions* 17 (1693): 596–610.

10. Ibid., p. 602.

11. Quoted in Samir Okasha, "Population Genetics," *Stanford Encyclopedia of Philosophy*, sec. 2, http://plato.stanford.edu/entries/population-genetics/ (accessed December 18, 2013).

12. G. H. Hardy, *A Mathematician's Apology* (Cambridge: Cambridge Canto Edition, 1992), p. 150.

13. George Bernard Shaw quoted in W. F. Bynum and R. Porter, eds., *Oxford Dictionary of Scientific Quotations* (Oxford: Oxford University Press, 2005), p. 547.

14. Geoffrey West and James Brown, "Life's Universal Scaling Laws," *Physics Today* (September 2004): 38.

CHAPTER 9: LIGHT

1. Samuel Johnson quoted in James Boswell, *Life of Johnson*, vol. 3 (Oxford: Talboys, 1826).

2. Samuel Johnson, *A Dictionary of the English Language* (London: 1755).

3. Ole Roemer in "A Demonstration Concerning the Motion of Light" (1676) in *Physical Thought from the Presocratics to the Quantum Physicists: An Anthology*, ed. S. Sambursky (New York: Pica, 1975).

4. William Wordsworth, "My Heart Leaps Up," in *Norton Anthology of Poetry*, 4th ed., ed. Margaret Ferguson, Mary Jo Salter, and John Stallworthy (New York: Norton, 1970), p. 728.

5. Isaac Newton, *Opticks* (New York: Dover, 1952), p. 175. A reprint of the original 4th ed. published in London in 1730.

6. John Keats, "Lamia," in *Poems of Science*, ed. John Heath-Stubbs and Phillips Salman (Harmondsworth, UK: Penguin, 1984), p. 197.

7. John Donne, "An Anatomy of the World," in Heath-Stubbs and Salman, *Poems of Science*, p. 79.

8. James Thomson, "To the Memory of Sir Isaac Newton," in Heath-Stubbs and Salman, *Poems of Science*, p. 138.

9. Newton, *Opticks*, pp. 1–2.

10. Christiaan Huygens, *Treatise on Light* (1690). Available in many sources, such as vol. 34 of *Great Books of the Western World* (Chicago: Encyclopedia Britannica, 1952), chap. 1.

11. G. B. Airy, "On the Diffraction of an Object-Glass with Circular Aperture," *Transactions of the Cambridge Philosophical Society* 5 (1835).

12. James Clerk Maxwell, *A Treatise on Electricity and Magnetism*, 3rd ed. (Oxford: Clarendon, 1904).

13. Heinrich Hertz, *Electric Waves*, as quoted in W. F. Bynum and R. Porter, eds., *Oxford Dictionary of Scientific Quotations* (Oxford: Oxford University Press, 2005), p. 279.

14. Albert Abraham Michelson, *Light Waves and Their Uses*, as quoted in Bynum and Porter, *Oxford Dictionary of Scientific Quotations*, p. 437.

15. Albert Einstein, "On a Heuristic Point of View about the Creation and Conversion of Light," *Annalen der Physik* 17 (1905): 549–60. See Anna Beck, trans., *The Collected Papers of Albert Einstein*, vol. 2 (Princeton, NJ: Princeton University Press, 1989), p. 87.

16. Einstein in a letter to a friend, quoted in A. Pais, *Subtle Is the Lord: The Science and Life of Albert Einstein* (Oxford: Oxford University Press, 1982), p. 30.

17. Robert A. Millikan in his book *The Electron*, as quoted in J. S. Rigden, *Einstein 1905: The Standard of Greatness* (Cambridge, MA: Harvard University Press, 2005), p. 37.

18. "The Nobel Prize in Physics 1921: Albert Einstein," Nobelprize.org, http://www.nobelprize.org/nobel_prizes/physics/laureates/1921/ (accessed April 7, 2015).

19. Arthur H. Compton, "A Quantum Theory of the Scattering of X-rays by Light Elements," *Physical Review* 21 (1923): 501.

20. Newton, *Opticks*, p. 339.

21. Einstein quoted in Bynum and Porter, *Oxford Dictionary of Scientific Quotations*, p. 279.

22. Einstein quoted in Pais, *Subtle Is the Lord*, p. 30.

23. Ibid., p. 414.

24. Einstein, letter to Michele Besso dated December 12, 1951, in Banesh Hoffman and Helen Dukas, eds., *Albert Einstein: Creator and Rebel* (New York: Viking, 1972).

CHAPTER 10: BUILDING BLOCKS

1. Lucretius, *De Rerum Natura*, written around 50 BCE and translated as *The Poem on Nature* by C. H. Sisson (Manchester, UK: Carcanet, 2006), bk. 1.

2. Richard Feynman, *The Feynman Lectures on Physics* (Reading, MA: Addison-Wesley, 1971), pp. 1–2.

3. James Clerk Maxwell, "Illustrations of the Dynamical Theory of Gases," *Philosophical Magazine* 9 (1860): 19–32.

4. Albert Einstein, "On the Movement of Small Particles Suspended in Stationary Liquids Required by the Molecular-Kinetic Theory of Heat," *Annalen der Physik* 17 (1905): 549–60. See Anna Beck, trans., *The Collected Papers of Albert Einstein*, vol. 2 (Princeton, NJ: Princeton University Press, 1989), p. 123.

5. Albert Einstein, "On the Theory of Brownian Motion," *Annalen der Physik* 19 (1906): 371–81. See Beck, *Collected Papers*, vol. 2, p. 180.

6. A. Pais, "Introducing Atoms and Their Nuclei," in *Twentieth Century Physics*, vol. 1, ed. L. M. Brown, A. Pais, and Sir Brian Pippard (Bristol, UK: Institute of Physics Publishing, 1995), p. 97.

7. "The Nobel Prize in Physics 1926: Jean Baptiste Perrin," Nobelprize.org, http://www.nobelprize.org/nobel_prizes/physics/laureates/1926/ (accessed April 7, 2015).

8. Max Born, "Einstein's Statistical Theories," in P. A. Schilpp, ed., *Albert Einstein: Philosopher-Scientist* (New York: Harper, 1959), p. 166.

9. G. Kirchhoff and R. Bunsen, "Chemische Analyse durch Spektralbeobachtungen," *Ostwalds Klassiker der exacten Wissenschaften*, no. 72 (1860), translated as "Chemical Analysis by Observation of the Spectrum," in *Physical Thought from the Presocratics to the Quantum Physicists: An Anthology*, ed. S. Sambursky (New York: Pica, 1975).

10. Johann Jakob Balmer, "Notiz über die Spectrallinien der Wasserstoffs," *Annalen der Physik und Chemie* 25 (1885): 80–85 (translated as "The Hydrogen Spectral Series," in *A Source Book in Physics*, ed. William Francis Maggie [Cambridge, MA: Harvard University Press, 1963], p. 360).

11. Words spoken to L. Rosenfeld, quoted in Niels Bohr, *On the Constitution of Atoms and Molecules*, with a forward by L. Rosenfeld (New York: Benjamin, 1963), p. 39.

12. Vistor Weisskopf quoted in W. F. Bynum and R. Porter, eds., *Oxford Dictionary of Scientific Quotations* (Oxford: Oxford University Press, 2005), p. 614.

13. Louis de Broglie in the 1929 Nobel lecture in physics (reproduced as "The Wave Nature of the Electron," in *Physical Thought from the Presocratics to the Quantum Physicists: An Anthology*, ed. S. Sambursky [New York: Pica, 1975], p. 514).

14. James Chadwick, "Possible Existence of a Neutron," letter to the editor, *Nature* 129 (1932): 312.

15. Ibid.

16. Lucretius, *De Rerum Natura*, translated as *Poem on Nature* by Sisson, bk. 1.

CHAPTER 11: NUCLEAR AND
PARTICLE PHYSICS

1. Richard Feynman, *QED: The Strange Theory of Light and Matter* (Princeton, NJ: Princeton University Press, 1985), pp. 7–8.

2. Ibid., p. 7.

3. Ibid.

4. Paul Dirac quoted in W. F. Bynum and R. Porter, eds., *Oxford Dictionary of Scientific Quotations* (Oxford: Oxford University Press, 2005), p. 179.

5. C. D. Ellis and W. A. Wooster, "The Average Energy of Disintegration of Radium E," *Proceedings of the Royal Society A*, cited in A. Pais, "Introducing Atoms and Their Nuclei," in *Twentieth Century Physics*, vol. 1, ed. L. M. Brown, A. Pais, and Sir Brian Pippard (Bristol, UK: Institute of Physics Publishing, 1995), p. 119.

6. Niels Bohr quoted in Pais, "Introducing Atoms and Their Nuclei," p. 127.

7. Wolfgang Pauli in a letter quoted in Laurie Brown, "The Idea of the Neutrino," *Physics Today* (September 1978): 27.

8. Fred Hoyle, *Astronomy and Cosmology: A Modern Course* (San Francisco: W. H. Freeman, 1975), p. 314.

9. Isaac Newton, when introducing his "axioms, or laws of motion," in *The Principia* (Amherst, NY: Prometheus Books, 1995), pp. 24–25.

10. Enrico Fermi quoted in Bynum and Porter, *Oxford Dictionary of Scientific Quotations*, p. 212.

11. Lucretius, *De Rerum Natura*, written around 50 BCE and translated as *The Poem on Nature* by C. H. Sisson (Manchester, UK: Carcanet, 2006), bk. 1.

12. Isaac Newton, *Opticks* (New York: Dover, 1952), p. 400. A reprint of the original 4th ed. published in London in 1730.

13. J. D. Cockcroft and E. T. S. Walton, "Disintegration of Lithium by Swift Protons," *Nature* 129 (1932): 649.

14. Sir Arthur Eddington, *The Internal Constitution of the Stars* (New York: Dover, 1959), p. 289. A reprint of the 1926 original.

15. Lord Kelvin, "On the Age of the Sun's Heat," *Macmillan's Magazine* 5 (1862): 388–93.

16. Eddington, *Internal Constitution of the Stars*, p. 291.

17. Fritz Houtermans quoted in Simon Singh, *Big Bang* (London: 4th Estate, 2004), p. 302.

18. "The Nobel Prize in Physics 1967: Hans Bethe," Nobelprize.org, http://www.nobelprize.org/nobel_prizes/physics/laureates/1967/ (accessed April 7, 2015).

19. Otto Hahn and Fritz Strassmann, "Concerning the Existence of Alkaline Earth Metals Resulting from Neutron Irradiation of Uranium," *Naturwissenschaften* 27 (January 1939): 12.

20. Otto Hahn and Fritz Strassmann, "Verification of the Creation of Radioactive Barium Isotopes from Uranium and Thorium by Neutron Irradiation; Identification of Additional Radioactive Fragments from Uranium Fission," *Naturwissenschaften* 27 (February 10, 1939): 95.

21. Lise Meitner and Otto R. Frisch, "Disintegration of Uranium by Neutrons: A New Type of Nuclear Reaction," *Nature* 143 (January 16, 1939): 239.

22. Robert Serber, *The Los Alamos Primer* (Berkeley: University of California Press, 1992).

23. Ibid.

24. Otto Frisch and Rudolf Peierls, "Peierls-Frisch Memorandum," reproduced in Serber, *Los Alamos Primer*, p. 82.

25. From the "Farm Hall transcripts" given in the book by Sir Charles Frank, *Operation Epsilon: The Farm Hall Transcripts* (Berkley: University of California Press, 1993), p. 71.

26. P. L. Rose, *Heisenberg and the Nazi Atomic Bomb Project: A Study in German Culture* (Berkeley: California University Press, 1998), p. 324.

CHAPTER 12: METHODS AND MOTION

1. Daniel Bernoulli, "Réflexions et éclaircissemens sur les nouvelles vibrations des cordes exposées dans les mémoires de l'Académie de 1747 & 1748," quoted in Morris Kline, *Mathematical Thought from Ancient to Modern Times* (New York: Oxford University Press, 1972), p. 509.

2. Lord Kelvin quoted in Alan Mackay, *A Dictionary of Scientific Quotations* (Bristol, UK: Adam Hilger, 1991), p. 239.

3. Lord Kelvin quoted in W. F. Bynum and R. Porter, eds., *Oxford Dictionary of Scientific Quotations* (Oxford: Oxford University Press, 2005), p. 582.

4. Ibid.

5. E. Fermi, J. Pasta, and S. Ulam, "Studies of Nonlinear Problems," *Los Alamos Document* LA-1940 (May 1955): 979.

6. Ibid.

7. Ibid.

8. Thierry Dauxois, "Fermi, Pasta, Ulam and a Mysterious Lady," *Physics Today* (January 2008): 57.

9. Robert M. May, "Simple Mathematical Models with Very Complicated Dynamics," *Nature* 261 (June 10, 1976): 467.

10. Mackay, *Dictionary of Scientific Quotations*, originally in Laplace, *Théorie Analatique des Probabilité*, vol. 3, *1812–1820*.

11. Robert May quoted in John Carey, ed., *The Faber Book of Science* (London: Faber, 1995), p. 504.

CHAPTER 13: EVALUATION

1. Quoted in Patricia Rife, *Lise Meitner and the Dawn of the Nuclear Age* (Boston: Birkhäuser, 1999), p. 21.

2. Emperor Napoleon Bonaparte quoted in W. F. Bynum and R. Porter, eds., *Oxford Dictionary of Scientific Quotations* (Oxford: Oxford University Press, 2005), p. 74.

3. Albert Einstein quoted in Bynum and Porter, *Oxford Dictionary of Scientific Quotations*, p. 201.

4. Sir Arthur Eddington quoted in Alan D. Mackay, *Dictionary of Scientific Quotations* (Bristol, UK: Adam Hilger, 1991), p. 79.

5. Fred Hoyle quoted in Mackay, *Dictionary of Scientific Quotations*, p. 123.

6. Robert Millikan quoted in Bynum and Porter, *Oxford Dictionary of Scientific Quotations*, p. 441.

BIBLIOGRAPHY

CHAPTER 1: INTRODUCTION

Baggott, J. *Farewell to Reality: How Modern Physics Has Betrayed the Search for Scientific Truth*. New York: Pegasus, 2013.

Cohen, I. Bernard. *The Triumph of Numbers*. New York: W. W. Norton, 2005.

Crease, R. P. *The Great Equations*. New York: Norton, 2008.

———. *The Prism and the Pendulum: The Ten Most Beautiful Experiments in Science*. New York: Random House, 2004.

Johnson, G. *The Ten Most Beautiful Experiments*. London: Vintage, 2009.

Malthus, T. R. *Essay on the Principle of Population as It Affects the Future Improvement of Society*. London: St. Paul's, 1798. (Many versions are available.)

Shamos, M. H. *Great Experiments in Physics*. New York: Holt, Rinehart and Winston, 1959.

CHAPTER 2: ANCIENT MATHEMATICS

Berggren, L., J. Borwein, and P. Borwein. *Pi: A Source Book*. New York: Springer-Verlag, 1997.

British Museum. "The Map of the World." http://www.britishmuseum.org/research/collection_online/collection_object_details/collection_image_gallery.aspx?assetId=404485&objectId=362000&partId=1 (accessed January 15, 2015).

Britton, J. P., C. Proust, and S. Shnider. "Plimpton 322: A Review and a Different Perspective." *Archive for the History of the Exact Sciences* 65 (2011): 519–66.

Brotton, Jerry. *A History of the World in Twelve Maps*. London: Penguin, 2013.

Calinger, R. *A Contextual History of Mathematics*. Upper Saddle River, NJ: Prentice Hall, 1999.

Chabert, Jean-Luc, ed. *A History of Algorithms*. Berlin: Springer-Verlag, 1999.

Dijksterhuis, E. J. *Archimedes*. Princeton, NJ: Princeton University Press, 1987.

George, Andrew. *The Epic of Gilgamesh*. London: Penguin Books, 1999.

Gill, A. *The Rise and Fall of Babylon*. London: Quercus, 2011.

Gillings, R. J. *Mathematics in the Time of the Pharaohs*. New York: Dover, 1972.

Hewath, Sir Thomas. *A History of Greek Mathematics*. Vol. 2. Oxford: The Clarendon, 1965.

Joseph, G. G. *The Crest of the Peacock: Non-European Roots of Mathematics*. London: Penguin, 1992.

Martzloff, J.-C. *A History of Chinese Mathematics*. Berlin: Springer-Verlag, 1997.

Neugebauer, O. *The Exact Science in Antiquity*. New York: Dover, 1959.

Robins, G., and C. Shute. *The Rhind Mathematical Papyrus*. London: British Museum Publications, 1987.

Robinson, A. *Writing and Script: A Very Short Introduction*. Oxford: Oxford University Press, 2009.

Robson, Eleanor. "Mathematics Education in an Old Babylonian Scribal School." In *The Oxford Handbook of the History of Mathematics*. Edited by Eleanor Robson and Jacqueline Stedall. Oxford: Oxford University Press, 2009.

———. "Neither Sherlock Holmes nor Babylon: A Reassessment of Plimpton 322." *Historia Mathematica* 28 (2001): 167–206.

———. "Tables and Tabular Formatting in Sumer, Babylonia, and Assyria, 2500BCE–50CE" In *The History of Mathematical Tables*. Edited by M. Campbell-Kelly, M. Croaken, R. Flood, and E. Robson. Oxford: Oxford University Press, 2003.

———. "Words and Pictures: New Light on Plimpton 322." *American Mathematical Monthly* 109 (2002): 105–20.

Stein, S. *Archimedes: What Did He Do besides Cry Eureka?* Washington, DC: Mathematical Association of America, 1999.

Yan, Li, and Du Shiran. *Chinese Mathematics: A Concise History*. Translated by J. N. Crossley and A. W.-C. Lun. Oxford: Clarendon Press, 1987.

CHAPTER 3: STEPS INTO MODERN MATHEMATICS

Berger, A., and T. P. Hill. "Benford's Law Strikes Back: No Simple Explanation in Sight for Mathematical Gem." *Mathematical Intelligencer* 33 (2011): 85–91.

Bruce, I. "The Agony and the Ecstasy—The Development of Logarithms by Henry Briggs." *Mathematical Gazette* 86 (2002): 216–27.

———. "Napier's Logarithms." *American Journal of Physics* 68 (2000): 148–54.

Calinger, R. *A Contextual History of Mathematics*. Upper Saddle River, NJ: Prentice Hall, 1999.

Carslaw, H. S. "The Discovery of Logarithms by Napier." *Mathematical Gazette* 8 (1915): 76–84, 115–19.

Chabert, Jean-Luc, ed. *A History of Algorithms*. Berlin: Springer-Verlag, 1999.

Conway, J. H., and R. K. Guy. *The Book of Numbers*. New York: Copernicus Springer-Verlag, 1996.

Devlin, K. *The Man of Numbers: Fibonacci's Arithmetic Revolution*. London: Bloomsbury, 2011.

Dunham, W. *Euler: The Master of Us All*. Washington, DC: Mathematical Association of America, 1999.

Edwards, A. W. F. *Pascal's Arithmetical Triangle*. London: Charles Griffin, 1987.

Euler, L., trans. *Introduction to Analysis of the Infinite*. By J. D. Blanton. New York: Springer-Verlag, 1988.

Gies, J., and F. Gies. *Leonardo of Pisa and the New Mathematics of the Middle Ages*. New York: Thomas Y. Crowell, 1969.

Goldstein, L. J. "A History of the Prime Number Theorem." *American Mathematical Monthly* 80 (1973): 599–615.

Goldstine, H. H. *A History of Numerical Analysis*. New York: Springer-Verlag, 1977.

Hill, T. P. "The First Digit Phenomenon." *American Scientist* 86 (1998): 358–63.

Horsburgh, E. M. *Modern Instruments and Methods of Calculation: A Handbook of the Napier Tercentenary Exhibition*. London: G. Bell and Sons, 1915.

Jagger, G. "The Making of Logarithmic Tables." In *History of Mathematical Tables*. Edited by M. Campbell-Kelly, M. Croaken, R. Flood, and E. Robson. Oxford: Oxford University Press, 2003.

Kline, M. "Euler and Infinite Series." *Mathematics Magazine* 56 (1983): 307–14.

Lehmer, D. N. "On the History of the Problem of Separating a Number into Its Prime Factors." *Scientific Monthly* 7 (1918): 227–34.

Sautoy, Marcus du. *Symmetry: A Journey into the Patterns of Nature*. New York: HarperCollins, 2008.

Sigler, L. E. *Fibonacci's* Liber Abaci: *A Translation into Modern English of Leonardo Pisano's* Book of Calculation. New York: Springer-Verlag, 2002.

Stillwell, J. *Mathematics and Its History*. New York: Springer-Verlag, 1989.

Tschinkel, Y. "About the Cover: On the Distribution of Primes—Gauss' Tables." *Bulletin of the American Mathematical Society* 43 (2005): 89–91.

Wells, D. *Prime Numbers: The Most Mysterious Figures in Math.* Hoboken, NJ: John Wiley, 2005.

Williams, M. R. *A History of Computing Technology.* Englewood Cliffs, NJ: Prentice Hall, 1985. (See ch. 2.)

CHAPTER 4: OUR WORLD

Auer, M., and M. Prior. "A New Era of Nuclear Test Verification." *Physics Today* (September 2014): 39–43.

Badash, L. "The Age-of-the-Earth Debate." *Scientific American* (August 1989): 78–83.

Bolt, B. A. *Earthquakes.* 5th ed. New York: W. H. Freeman, 2004.

Brush, S. G. "Discovery of the Earth's Core." *American Journal of Physics* 48 (1980): 705–24.

Buffett, B. "Earth's Enigmatic Inner Core." *Physics Today* 66 (November 2013): 37–41.

Bullen, K. E., and B. A. Bolt. *An Introduction to the Theory of Seismology.* 4th ed. Cambridge: Cambridge University Press, 1985.

Cartright, D. E. *Tides: A Scientific History.* Cambridge, UK: Cambridge University Press, 1999.

Center for Operational Oceanographic Products and Services. "Harmonic Analysis." http://tidesandcurrents.noaa.gov/harmonic.html (accessed March 20, 2014).

Doodson, A. T. "The Development of Tidal Theory." *Journal of Navigation* 6 (1953): 177–83.

Dutka, J. "Eratosthenes' Measurement of the Earth Reconsidered." *Archive for the History of Exact Sciences* 46 (1993): 55–66.

England, P. C., P. Molnar, and F. M. Richter. "Kelvin, Perry and the Age of the Earth." *American Scientist* 95 (2007): 342–49.

Galilei, Galileo. *Two New Sciences:* Translated by Stillman Drake. Madison: University of Wisconsin Press, 1974. (Many versions of this book are available.)

Herak, Marijan. "Mohorovičić Discontinuity." *Andrija Mohorovičić's Memorial Rooms.* http://www.gfz.hr/sobe-en/discontinuity.htm (accessed March 12, 2014).

Jarchow, C. M., and G. A. Thompson. "The Nature of the Mohorovičić Disconti-nuity." *Annual Review of Earth and Planetary Science* 17 (1989): 475–506.

Kölbl-Ebert, M. "Inge Lehmann's Paper: 'P' (1936)." *Classic Papers in the History of Geology* 24 (2001): 262–67.

Lehmann, Inge. "Seismology in the Days of Old." *Eos Transactions American Geophysical Union* 68 (1987): 33–35.

Nicastro, N. *Circumference: Eratosthenes and the Ancient Quest to Measure the Globe.* New York: St. Martin's, 2008.

Oldham, R. D. "The Constitution of the Interior of the Earth as Revealed by Earth-quakes." *Quarterly Journal of the Geological Society* 62 (1906): 456–75.

———. "On the Propagation of Earthquake Motion to Great Distances." *Philo-sophical Transactions of the Royal Society of London, Series A* 194 (1900): 135–74.

Oldroyd, D. R. *Thinking about the Earth: A History of Ideas in Geology.* Cam-bridge, MA: Harvard University Press, 1996.

Olson, P. "The Geodynamo's Unique Longevity." *Physics Today* 66 (November 2013): 30–35.

Parker, B. "The Tide Predictions for D-Day." *Physics Today* 64 (September 2011): 35–40.

Perry, J. "On the Age of the Earth." *Nature* 51 (1895): 224–27, 341–42, 582–85.

Phillips, T. "Fourier Analysis of Ocean Tides." http://www.ams.org/samplings/feature-column/fcarc-tidesii1 (accessed March 20, 2014).

Physics Update. "Iron's Structure at Earth's core." *Physics Today* (December 2010): 26.

Rousseau, C. "How Inge Lehmann Discovered the Inner Core of the Earth." *College Mathematics Journal* 44 (2013): 399–408.

Ruddock, I. "Lord Kelvin, Science and Religion." *Physics World* (February 2004): 56.

Smith, C., and M. Norton Wise. *Energy and Empire: A Biographical Study of Lord Kelvin.* Cambridge: Cambridge University Press, 1989.

Thomson, Sir William (Lord Kelvin). "On the Age of the Earth." *Nature* 51 (1895): 438–40.

———. "On the Secular Cooling of the Earth." *Transactions of the Royal Society of Edinburgh* 23 (1862): 157–69.

———. "The Tides." Evening Lecture to the British Association at the South-ampton Meeting, Friday, August 25, 1882.

Thomson, Sir William (Lord Kelvin), and P. G. Tait. *Principles of Mechanics and Dynamics (*formerly titled *Treatise on Natural Philosophy).* New York: Dover, 1962. (A republication of the last revised edition of 1912.)

Umino, S., K. Nealson, and B. Wood. "Drilling to Earth's Mantle." *Physics Today* 66 (August 2013): 36–41.

Wainwright, J., and M. Mulligan. *Environmental Modelling: Finding Simplicity in Complexity*. Chichester, UK: Wiley-Blackwell, 2013.

Weintraub, D. A. *How Old Is the Universe?* Princeton, NJ: Princeton University Press, 2011.

Wikipedia. "Tide Predicting Machine." http://en.wikipedia.org/wiki/Tide -predicting_machine (accessed March 18, 2014).

CHAPTER 5: THE SOLAR SYSTEM: THE FIRST MATHEMATICAL MODELS

Beer, A., and P. Beer, eds. *Kepler, Four Hundred Years: Proceedings of Conferences Held in Honour of Johannes Kepler*. Oxford: Pergamon, 1975.

Chabert, Jean-Luc, ed. *A History of Algorithms*. Berlin: Springer-Verlag, 1999.

Gingerich, O. "The Great Martian Catastrophe and How Kepler Fixed It." *Physics Today* 64 (September 2011): 50–54.

———. "Johannes Kepler." in *The General History of Astronomy*. Vol. 2, *Planetary Astronomy from the Renaissance to the Rise of Astrophysics Part A: Tycho Brahe to Newton*. Edited by R. Taton and C. Wilson. Cambridge: Cambridge University Press, 1989.

———. "Kepler." In *Dictionary of Scientific Biography*. Edited by C. C. Gillespie. New York: Charles Scribner's Sons, 1975.

Hanson, N. R. "The Mathematical Power of Epicycle Astronomy." *Isis* 51 (1960): 150–58.

Koyré, A. *The Astronomical Revolution*. London: Routledge, 2009.

Linton, C. M. *From Eudoxus to Einstein: A History of Mathematical Astronomy*. Cambridge: Cambridge University Press, 2004.

Pedersen, O. "Scientific Accounts of the Universe from Antiquity to Kepler." *European Review* 2 (1994): 125–40.

Ptolemy. *Almagest*. Translated and annotated by G. J. Toomer. London: Duckworth, 1984.

Thurston, H. *Early Astronomy*. New York: Springer, 1996.

Toomer, G. J. "Ptolemy." In *Dictionary of Scientific Biography*. Edited by C. C. Gillespie. New York: Charles Scribner's Sons, 1975.

Van Brummelen, G. *The Mathematics of the Heavens and the Earth: The Early History of Trigonometry*. Princeton, NJ: Princeton University Press, 2009.

CHAPTER 6: THE SOLAR SYSTEM: INTO THE MODERN ERA

Cohen, I. B. "Newton's Determination of the Masses and Densities of the Sun, Jupiter, Saturn, and the Earth." *Archive for the History of Exact Sciences* 53 (1998): 83–95.

Einstein, A. "The Foundations of the General Theory of Relativity." In *The Principle of Relativity*. New York: Dover, 1952. (A translation by W. Perrett and G. B. Jeffery of Einstein's original, "Die Grundlage der allgemeinen Relativitätstheorie," *Annalen der Physik* 49 [1916]: 769.)

Fernie, J. D. "In Pursuit of Vulcan." *American Scientist* 82 (1994): 412–15.

Grier, D. A. *When Computers Were Human*. Princeton, NJ: Princeton University Press, 2005.

Grosser, M. *The Discovery of Neptune*. New York: Dover, 1979.

Halley, E. "A New Method of Determining the Parallax of the Sun." *Philosophical Transactions* 29 (1716): 454–56. Translation from the Latin available at http://eclipse.gsfc.nasa.gov/transit/HalleyParallax.html (accessed April 12, 2014).

Hanson, N. R. "Leverrier: The Zenith and Nadir of Newtonian Mechanics." *Isis* 53 (1962): 359–78.

Kent, D. "The Curious Aftermath of Neptune's Discovery." *Physics Today* 64 (December 2011): 46–51.

Linton, C. M. *From Eudoxus to Einstein: A History of Mathematical Astronomy*. Cambridge: Cambridge University Press, 2004.

Lomb, N. *Transit of Venus 1631 to the Present*. Sydney: NewSouth, 2011.

Marsden, B. G. "Eighteenth and Nineteenth Century Developments in the Theory and Practice of Orbit Determination." In *The General History of Astronomy*. Vol. 2, *Planetary Astronomy from the Renaissance to the Rise of Astrophysics Part B: The Eighteenth and Nineteenth Centuries*. Edited by R. Taton and C. Wilson. Cambridge: Cambridge University Press, 1989.

Morando, B. "The Golden Age of Celestial Mechanics." In *The General History of Astronomy*. Vol. 2, *Planetary Astronomy from the Renaissance to the Rise of Astrophysics Part B: The Eighteenth and Nineteenth Centuries*. Edited by R. Taton and C. Wilson. Cambridge: Cambridge University Press, 1989.

Pais, A. *Subtle Is the Lord: The Science and Life of Albert Einstein*. Oxford: Oxford University Press, 1982.

Park, D. *Classical Dynamics and Its Quantum Analogues*. Berlin: Springer-Verlag, 1979.

Pask, C. *Magnificent Principia: Exploring Isaac Newton's Masterpiece*. Amherst, NY: Prometheus Books, 2013.

Pedersen, O. "Scientific Accounts of the Universe from Antiquity to Kepler." *European Review* 2 (1994): 125–40.

Phillips, L. *The Transit of Venus*. http://brightstartutors.com/blog/2012/the-transit-of-venus (accessed March 4, 2014)

Ronan, C. A. "Edmond Halley." In *Dictionary of Scientific Biography*. Edited by C. C. Gillespie. New York: Charles Scribner's Sons, 1975.

———. *Edmond Halley: Genius in Eclipse*. London: MacDonald, 1970.

Smith, R. W. "The Cambridge Network in Action: The Discovery of Neptune." *Isis* 80 (1989): 395–422.

Standage, T. *The Neptune File*. London: Allen Lane Penguin, 2000.

Van Helden, A. "Measuring Solar Parallax: The Venus Transits of 1761 and 1769 and Their Nineteenth Century Sequels." In *The General History of Astronomy*. Vol. 2, *Planetary Astronomy from the Renaissance to the Rise of Astrophysics Part B: The Eighteenth and Nineteenth Centuries*. Edited by R. Taton and C. Wilson. Cambridge: Cambridge University Press, 1989.

———. *Measuring the Universe: Cosmic Dimensions from Aristarchus to Halley*. Chicago: University of Chicago Press, 1985.

Waff, C. B. "Predicting the Mid-Eighteenth-Century Return of Halley's Comet." In *The General History of Astronomy*. Vol. 2, *Planetary Astronomy from the Renaissance to the Rise of Astrophysics Part B: The Eighteenth and Nineteenth Centuries*. Edited by R. Taton and C. Wilson. Cambridge: Cambridge University Press, 1989.

Woolf, H. *The Transits of Venus: A Study of Eighteenth Century Science*. Princeton, NJ: Princeton University Press, 1959.

CHAPTER 7: THE UNIVERSE

Barger, V. D., and M. G. Olsson. *Classical Mechanics: A Modern Perspective*. 2nd ed. New York: McGraw Hill, 1995.

Burbidge, E. M., and G. R. Burbidge. "The Masses of Galaxies." Ch. 3 in *Galaxies and the Universe*. Edited by A. Sandage, M. Sandage, and J. Kristian. Chicago: University of Chicago Press, 1975.

Burbidge, E. M., G. R. Burbidge, W. A. Fowler, and F. Hoyle. "Synthesis of the Elements in Stars." *Reviews of Modern Physics* 29 (1957): 547–650.

Caldwell, R., and M. Kamionkowski. "Dark Matter and Dark Energy." *Nature* 458 (2009): 587–89.

Coles, P. *Cosmology: A Very Short Introduction*. Oxford: Oxford University Press, 2001.

Davies, Paul. *The Goldilocks Enigma: Why is the Universe Just Right for Life?* Boston: Houghton Mifflin, 2008.

Earman, J. "Lambda: The Constant that Refuses to Die." *Archive for History of the Exact Sciences* 55 (2001): 189–220.

Einasto, J., A. Kaasik, and E. Saar. "Dynamic Evidence on the Massive Coronas of Galaxies." *Nature* 250 (1974): 309–10.

Freedman, W. L., and B. F. Madore. "The Hubble Constant." *Annual Review of Astronomy and Astrophysics* 48 (2010): 673–710.

Gardner, Martin. "WAP, SAP, PAP and FAP." In *The Night Is Large: Collected Essays 1938–1995*. New York: St. Martin's, 1996.

Hansen, J. R. *Enchanted Rendezvous: John C. Houbolt and the Genesis of the Lunar-Orbit Rendezvous Concept*. Washington, DC: NASA Monographs in Aerospace History Series 4, 1995.

Harrison, E. R. *Cosmology: The Science of the Universe*. 2nd ed. Cambridge: Cambridge University Press, 2000.

———. *Darkness at Night: A Riddle of the Universe*. Cambridge, MA: Harvard University Press, 1987.

———. "The Dark Night-Sky Riddle: A "Paradox that Resisted Solution." *Science* 226 (1984): 941–45.

———. "The Dark Sky Paradox." *American Journal of Physics* 45 (1977): 119–24.

———. "Kelvin on an Old Celebrated Hypothesis." *Nature* 322 (1986): 417–18.

Hoyle, Fred. *Astronomy and Cosmology: A Modern Course*. San Francisco: W. H. Freeman, 1975.

———. *Home is Where the Wind Blows: Chapters from a Cosmologist's Life*. Mill Valley, CA: University Science Books, 1994.

———. "On Nuclear Reactions Occurring in Very Hot Stars. The Synthesis of Elements from Carbon to Nickel." Supplement 1. *Astrophysics Journal* 1 (1954): 121–46.

Hubble, E. "A Relation between the Distance and Radial Velocity among Extra-Galactic Nebulae." *Proceedings of the National Academy of Sciences* 15 (1929): 168–73.

Irion, Robert. "The Bright Face behind the Dark Side of Galaxies." *Science* 295 (2002): 960–61.

Konopliv, A. S., et al. "Improved Gravity Field of the Moon from Lunar Prospector." *Science* 281 (1998): 1476–80.

Kragh, H. "An Anthropic Myth: Fred Hoyle's Carbon-12 Resonance Level." *Archive for the History of Exact Sciences* 64 (2010): 721–51.

Lambourne, R. J. A. *Relativity, Gravitation and Cosmology*. Cambridge: Cambridge University Press, 2010.

Lee, T. E. *Bizarre Lunar Orbits*. NASA Science. http://science.nasa.gov/science-news/science-at-nasa/2006 (accessed May 22, 2014.)

Livio, M. "Mystery of the Missing Text Solved." *Nature* 479 (2011): 171–73.

Ostriker, J. P., P. J. E. Peebles, and A. Yahil. "The Size and Mass of Galaxies, and the Mass of the Universe." *The Astrophysical Journal* 193 (1974): L1–L4.

Roos, M. *Introduction to Cosmology*. 3rd ed. Chichester, UK: Wiley, 2003.

Roy, A. E. *Orbital Motion*. 4th ed. Bristol, UK: Institute of Physics Publishing, 2005.

Rubin, Vera C., and W. Kent Ford. "Rotation of the Andromeda Nebula from a Spectroscopic Survey of Emission Regions." *Astrophysical Journal* 159 (1970): 379–92.

Rubin, Vera C. "Dark Matter in Spiral Galaxies." *Scientific American* (1983): 96–108.

———. "The Rotation of Spiral Galaxies." *Science* 220 (1983): 1339–44.

———. "Seeing Dark Matter in the Andromeda Galaxy." *Physics Today* (December 2006): 8–9.

Van Allen, J. A. "Gravitational Assist in Celestial Mechanics—A Tutorial." *American Journal of Physics* 71 (2003): 448–51.

Way, M., and H. Nussbaumer. "Lemaitre's Hubble Relationship." *Physics Today*. Letters. (August 2011): 8.

Weinberg, S. *The First Three Minutes*. 2nd ed. New York: Basic Books, 1988.

CHAPTER 8: ABOUT US

Banavar, J. R., et al. "A General Basis for Quarter-Power Scaling in Animals." *Proceedings of the National Academy of Sciences* 107 (September 2010): 15816–20.

Brown, J. H., and G. B. West, eds. *Scaling in Biology*. Oxford: Oxford University Press, 2000. (This book contains many papers given at a symposium on scaling.)

Burton, R. F. *Biology by Numbers: An Encouragement to Quantitative Thinking*. Cambridge: Cambridge University Press, 1998.

Ciecka, J. E. "Edmond Halley's Life Table and Its Uses." *Journal of Legal Economics* 15 (2008): 65–74.

Cohen, I. Bernard. *The Triumph of Numbers*. New York: W. W. Norton, 2005.

Cormack, A. M. "Early Two-Dimensional Reconstruction and Recent Topics Stemming from It." *Science* 209 (1980): 1482–86.

———. "Representation of a Function by Its Line Integrals, with Some Radiological Applications." *Journal of Applied Physics* 34 (1963): 2722–27.

———. "Representation of a Function by Its Line Integrals, with Some Radiological Applications. II" *Journal of Applied Physics* 35 (1964): 2908–13.

Epstein, C. L. *Introduction to the Mathematics of Medical Imaging*. Upper Saddle River, NJ: Pearson Education, 2003.

Glazier, D. S. "The 3/4-Power Law is Not Universal: Evolution of Isometric, Ontogenetic Metabolic Scaling in Pelagic Animals." *BioScience* 56 (2006): 325–32.

———. "A Unifying Explanation for Diverse Metabolic Scaling in Animals and Plants." *Biological Reviews* 85 (2010): 111–38.

Gordon, R., G. T. Herman, and S. A. Johnson. "Image Reconstruction from Projections." *Scientific American* 233 (October 1975): 56–68.

Gregory, Andrew. *Harvey's Heart: The Discovery of Blood Circulation*. Duxford, UK: Icon Books, 2001.

Hald, A. "The Early History of Life Insurance Mathematics." Ch. 9 in *A History of Probability and Statistics and Their Application before 1750*. New York: John Wiley, 1990.

Halley, E. "An Estimate of the Degrees of the Mortality of Mankind, Drawn from Curious Tables of the Births and Funerals at the City of Breslaw; with an Attempt to Ascertain the Price of Annuities upon Lives." *Philosophical Transactions* 17 (1693): 596–610.

Hardy, G. H. *A Mathematician's Apology*. Cambridge: Cambridge Canto Edition, 1992.

———. "Mendelian Proportions in a Mixed Population." *Science* 28 (1908): 49–50.

Harvey, W. *An Anatomical Disquisition on the Motion of the Heart and Blood in Animals, A Translation of the 1628 Original* Exercitatio Anatomica de Motu Cordis et Sanguinis in Animalibus. Chicago: Great Books of the Western World, Encyclopedia Britannica, 1990.

Kac, A. C. *Principles of Computerized Tomographic Imaging*. New York: IEEE, 1988.

Kopf, E. W. "The Early History of the Annuity." http://www.casact.org/pubs/proceed/proceed26/26225.pdf (accessed June 2, 2014).

Lewin, C., and M. De Valois. "History of Actuarial Tables." In *The History of Mathematical Tables*. Edited by M. Campbell-Kelly, M. Croaken, R. Flood, and E. Robson. Oxford: Oxford University Press, 2003.

Newman, J. R. *The World of Mathematics*. London: George Allen and Unwin, 1961. Vol. 3 contains excerpts from the work of Graunt and Halley.

Okasha, S. "Population Genetics." *Stanford Encyclopedia of Philosophy*. http://plato.stanford.edu/entries/population-genetics/ (accessed December 18, 2013).

Prange, H. D., J. F. Anderson, and H. Rahn. "Scaling of Skeletal Mass to Body Mass in Birds and Mammals." *American Naturalist* 113 (1979): 103–22.

Savage, V. M., et al. "The Predominance of Quarter-Power Scaling in Biology." *Functional Ecology* 18 (2004): 257–82.

Schmidt-Nielsen, K. *How Animals Work*. Cambridge: Cambridge University Press, 1972.

———. *Scaling: Why Is Animal Size So Important?* Cambridge: Cambridge University Press, 1984.

Shepp, L. A., and J. B. Kruskal. "Computerized Tomography: The New Medical X-ray Technology." *American Mathematical Monthly* 85 (1978): 420–39.

Spence. A. J. "Scaling in Biology." *Current Biology* 19 (January 27, 2009): R57–R61.

Vogel, S. *Life's Devices*. Princeton, NJ: Princeton University Press, 1988.

Webb, S. *From the Watching of Shadows: The Origins of Radiological Tomography*. Bristol, UK: Adam Hilger, 1990.

West, J. B., and J. H. Brown. "Life's Universal Scaling Laws." *Physics Today* (September 2004): 36–42.

Whitfield, J. *In the Beat of a Heart: Life, Energy and the Unity of Nature*. Washington, DC: Joseph Henry, 2006.

CHAPTER 9: LIGHT

Airy, G. B. "On the Diffraction of an Object-Glass with Circular Aperture." *Transactions of the Cambridge Philosophical Society* 5 (1835): 283–91.

Boyer, C. B. *The Rainbow: From Myth to Mathematics*. Princeton, NJ: Princeton University Press, 1989.

Brush, S. G. "Early Estimates of the Velocity of Light." *Isis* 33 (March 1941): 24–40.

———. "Prediction and Theory Evaluation: The Case of Light Bending." *Science* 246 (December 1, 1989): 1124–29.

Cantor, G. N. *Optics After Newton*. Manchester, UK: Manchester University Press, 1983.

Casselman, B. *The Mathematics of Rainbows*. American Mathematical Society Monthly Essays. http://www.ams.org/samplings/feature-column/fcarc-rainbows (accessed June 27, 2014).

Cohen, I. B. "Roemer and the First Determination of the Velocity of Light (1676)." *Isis* 31 (April 1940): 327–79. This paper contains a facsimile of Roemer's original 1676 paper.

Daukantas, P. "Ole Roemer and the Speed of Light." *Optics and Photonic News* (July 2009): 43–47.

Denny, M., and A. McFadzean. *Engineering Animals: How Life Works*. Cambridge, MA: Belknap, 2011. (See ch. 10 for photoreceptors and image sampling.)

Eddington, A. *Space, Time and Gravitation: An Outline of the General Theory of Relativity*. New York: Harper, 1959. (A reprint of the original 1920 edition.)

Einstein, A. "On a Heuristic Point of View about the Creation and Conversion of Light." *Annalen der Physik* 17 (1905): 132–48. (This paper is reprinted in various books and in Einstein's collected papers.)

Feynman, R. P. *The Feynman Lectures on Physics*. Vol. 2, *Mainly Electromagnetism and Matter*. Reading, MA: Addison-Wesley, 1971.

———. *QED: The Strange Theory of Light and Matter*. Princeton, NJ: Princeton University Press, 1985.

Gamow, G. *Thirty Years That Shook Physics*. New York: Dover, 1966.

Harman, P. M. *The Natural Philosophy of James Clerk Maxwell*. Cambridge: Cambridge University Press, 1998.

Heath-Stubbs, J., and P. Salman, eds. *Poems of Science*. Harmondsworth, UK: Penguin, 1984.

Heavens, O. S., and R. W. Ditchburn. *Insight into Optics*. Chichester, UK: John Wiley, 1991.

Huygens, C. *Treatise on Light*. 1690. (Available in many sources, such as vol. 34 of *Great Books of the Western World* [Chicago: Encyclopedia Britannica, 1952], chap. 1.)

Kennefick, D. "Testing Relativity from the 1919 Eclipse—A Question of Bias." *Physics Today* (March 2009): 37–42.

Lambourne, R. J. A. *Relativity, Gravitation and Cosmology*. Cambridge: Cambridge University Press, 2010.

Lipson, S. G., and H. Lipson. *Optical Physics*. Cambridge: Cambridge University Press, 1969.

Mahon, B. *The Man Who Changed Everything: The Life of James Clerk Maxwell*. Chichester, UK: Wiley, 2004.

Maxwell, J. C. *A Treatise on Electricity and Magnetism*. 3rd ed. Oxford: Clarendon, 1904.

Newton, Isaac. *Opticks*. New York: Dover, 1952. (A reprint of the original 4th ed. published in London in 1730.)

Nicolson, Marjorie Hope. *Newton Demands the Muse*. Princeton, NJ: Princeton University Press, 1966.

Nussenzveig, H. M. "The Theory of the Rainbow." *Atmospheric Phenomena: Readings from Scientific American*. San Francisco: W. H. Freeman, 1980.

Pais, A. *Subtle Is the Lord: The Science and Life of Albert Einstein*. Oxford: Oxford University Press, 1982.

Rigden, J. S. *Einstein 1905: The Standard of Greatness*. Cambridge, MA: Harvard University Press, 2005.

Robinson, A. *The Last Man Who Knew Everything*. New York: Pi Press, 2006.

Roemer, O. "A Demonstration Showing the Motion of Light." In *Physical Thought from the Presocratics to the Quantum Physicists: An Anthology*. Edited by S. Sambursky. New York: Pica, 1975.

Roychoudhuri, C., and R. Roy. eds. "The Nature of Light. What Is a Photon?" *Trends*, supplement of *Optics and Photonics News*. (October 2003): S1–S35.

Segrè, E. *From Falling Bodies to Radio Waves*. New York: W. H. Freeman, 1984. (See ch. 4, "Electricity: From Thunder to Motors and Waves.")

Soldner, Johann Georg von. "On the Deflection of a Light Ray from Its Rectilinear Motion, by the Attraction of a Celestial Body at Which It Nearly Passes By." *Berliner Astronomisches Jahrbuch* (1804): 161–72. (Translations are readily available; for example, see S. L. Jaki in *Foundations of Physics* 8 (1978): 927–50.)

Wandell, B. A. *Foundations of Vision*. Sunderland, MA: Sinauer, 1995.

Weinberg, S. *Gravitation and Cosmology: Principles and Applications of the General Theory of Relativity*. New York: John Wiley, 1972.

Will, C. M. *Was Einstein Right? Putting General Relativity to the Test*. New York: Basic Books, 1993.

———. "Henry Cavendish, Johann von Soldner, and the Deflection of Light." *American Journal of Physics* 56 (May 1988): 413–15.

Williams, B., ed. *Compton Scattering*. London: McGraw-Hill, 1977. (The paper "History" by R. H. Stuewer and M. J. Cooper gives a good introduction to Compton scattering.)

Williamson, S. J., and H. Z. Cummins. *Light and Color in Nature and Art*. New York: John Wiley, 1983.

CHAPTER 10: BUILDING BLOCKS

Balmer, J. J. "Notiz über die Spectrallinien des Wasserstoffs." *Annalen der Physik und Chemie* 25 (1885): 80. Translated as "The Hydrogen Spectral Series" in *A Source Book in Physics*. Edited by W. F. Magie. Cambridge, MA: Harvard University Press, 1963.

Bernstein, J. "Einstein and the Existence of Atoms." *American Journal of Physics* 74 (October 2006): 863–72.

De Broglie, Louis. "The Wave Nature of the Electron." 1929 Nobel Prize acceptance speech, reprinted in *Physical Thought from the Presocratics to the Quantum Physicists: An Anthology*. Edited by S. Sambursky. New York: Pica, 1975.

Chadwick, J. Nobel Prize in Physics Award Address, 1935. Reprinted in *The World of Physics*. Vol. 2. Edited by F. H. Weaver. New York: Simon and Schuster, 1987.

———. "Possible Existence of a Neutron." Letter to the Editor in *Nature* 129 (1932): 312.

Clark, C. W., and J. Reader. "The Optical Discovery of Deuterium." *Optics and Photonics News* 23 (May 2012): 36–41.

Einstein, A. *The Collected Papers of Albert Einstein*. Translated by Anna Beck. Princeton, NJ: Princeton University Press, 1989. (Vol. 2 contains the relevant Brownian–motion papers.)

Fayer, M. D. *Absolutely Small: How Quantum Theory Explains Our Everyday World*. New York: AMACOM, 2010.

Feynman, R. P. *The Feynman Lectures on Physics*. Reading, MA: Addison-Wesley, 1971.

Gamow, George. *Thirty Years That Shook Physics*. New York: Dover, 1966.

Holton, G., and S. G. Brush. *Introduction to Concepts and Theories in Physical Science*. 2nd ed. Reading, MA: Addison-Wesley, 1973.

Kirchhoff, G., and R. Bunsen. "Chemische Analyse durch Spektralbeobach-tungen." *Ostwalds Klassiker der exacten Wissenschaften*, no. 72 (1860). Translated as "Chemical Analysis by Observation of the Spectrum" in *Physical Thought from the Presocratics to the Quantum Physicists: An Anthology*. Edited by S. Sambursky. New York: Pica, 1975.

Lucretius. *De Rerum Natura*. Written around 50 BCE. Translated as *The Poem on Nature* by C. H. Sisson. Manchester: Carcanet, 2006.

Newburgh, R., J. Peidle, and W. Ruekner. "Einstein, Perrin, and the Reality of Atoms: 1905 Revisited." *American Journal of Physics* 74 (June 2006): 478–81.

Pais, A. "Introducing Atoms and Their Nuclei." In *Twentieth Century Physics*. Vol. 1. Edited by L. M. Brown, A. Pais, and Sir Brian Pippard. Bristol, UK: Institute of Physics Publishing, 1995.

Pask, C. *Math for the Frightened*. Amherst, NY: Prometheus Books, 2011. (See ch. 13.)

Perrin, M. Jean. *Brownian Movement and Molecular Reality*. London: Taylor and Francis, 1910. (A translation by F. Soddy of Perrin's article in *Annales de Chimie et de Physique*, 8th series [September 1909].)

Rechenberg, H. "Quanta and Quantum Mechanics," in *Twentieth Century Physics*. Vol. 1. Edited by L. M. Brown, A Pais, and Sir Brian Pippard. Bristol, UK: Institute of Physics Publishing, 1995.

Rigden, J. S. *Einstein 1905: The Standard of Greatness*. Cambridge, MA: Harvard University Press, 2005. (See the "'Seeing' Atoms" chapter.)

Ter Haar, D. *The Old Quantum Theory*. Oxford: Pergamon, 1967.

Wallace, P. R. *Paradox Lost: Images of the Quantum*. New York: Springer, 1996.

CHAPTER 11: NUCLEAR AND
PARTICLE PHYSICS

Ashby, N. "Relativity and the Global Positioning System." *Physics Today* (May 2002): 41–47.

Atkinson, R., and F. Houtermans. "Aufbaumölichkeit in Sternen." *Zeitschrift für Physik* 54 (1929): 656–65.

Bahcall, J. N. "Solving the Mystery of the Missing Neutrinos." Nobelprize.org. http://www.nobelprize.org/nobel_prizes/themes/bahcall/ (accessed August 24, 2014).

Bethe, H. A. "Energy Production in Stars." *Physical Review* 55 (1939): 434–56.

Brown, L. "The Idea of the Neutrino." *Physics Today* (September 1978): 23–28.

Brush, S. G. *Kinetic Theory*. Vol. 1, *The Nature of Gases and of Heat*. Oxford: Pergamon Press, 1965.

Cockcroft, J., and E. Walton. "Disintegration of Lithium by Swift Protons." *Nature* 129 (1932).

———. "Experiments with High Velocity Positive Ions. II – The Disintegration of Elements by High Velocity Protons." *Proceedings of the Royal Society A* 227 (1932): 229–36.

Dirac, P. A. M. Nobel Prize in Physics, Award Address, 1933. In *The World of Physics*. Vol. 2. Edited by J. H. Weaver. New York: Simon and Schuster, 1987.

———. *The Principles of Quantum Mechanics*. 4th ed. Oxford: Clarendon, 1958.

Dunlap, R. A. *The Physics of Nuclei and Particles*. Toronto: Thomson Brooks/ Cole, 2004.

Dürr, S., et al. "Ab-Initio Determination of Light Hadron Masses." *Science* 322, no. 5905 (November 21, 2008): 1224–27.

Eddington, A. S. *The Internal Constitution of Stars*. New York: Dover, 1959. (A reprint of the 1926 original. The relevant part is also reprinted in the book by Kilmister, noted below.)

Farmelo, G. *The Strangest Man: The Hidden Life of Paul Dirac, Mystic of the Atom*. New York: Basic Books, 2009.

Fayer, M. D. *Absolutely Small: How Quantum Theory Explains Our Everyday World*. New York: AMACOM, 2010.

Feynman, R. P. *QED: The Strange Theory of Light and Matter*. Princeton, NJ: Princeton University Press, 1985.

———. "Space-Time Approach to Quantum Electrodynamics." *Physical Review* 76 (1949): 769–89.

Frank, Sir Charles. *Operation Epsilon: The Farm Hall Transcripts*. Berkley: University of California Press, 1993.

Gabrielse, G. "The Standard Model's Greatest Triumph." *Physics Today* (December 2013): 64– 65.

Graetzer, H. G., and D. L. Anderson. *The Discovery of Nuclear Fission: A Documentary History*. New York: Van Nostrand, 1971. (A wonderful collection of original papers and commentary on them.)

Greenberg, A. J., et al. "Charged-Pion Lifetime and a Limit on a Fundamental Length." *Physical Review Letters* 23 (1969): 1267–70.

Halzen, F., and S. R. Klein. "Astronomy and Astrophysics with Neutrinos." *Physics Today* (May 2008): 29–35.

Kelvin, Lord. "On the Age of the Sun's Heat." *Macmillan's Magazine* 5 (1862): 388–93.

Kilmister, C. W. *Men of Physics: Sir Arthur Eddington*. Oxford: Pergamon, 1966.

King, A. *Stars: A Very Short Introduction*. Oxford: Oxford University Press, 2012.

Lambourne, R. J. A. *Relativity, Gravitation and Cosmology*. Cambridge: Cambridge University Press, 2010.

Lang, K. R., and O. Gingerich, eds. *A Sourcebook in Astronomy and Astrophysics, 1900–1975*. Cambridge, MA: Harvard University Press, 1975.

Logan, J. "The Critical Mass." *American Scientist* 84 (May–June 1996): 263–77.

McCracken, G., and P. Stott. *Fusion: The Energy of the Universe*. Amsterdam: Elsevier, 2005.

Mermin, D. *It's About Time: Understanding Einstein's Relativity*. Princeton, NJ: Princeton University Press, 2005.

National Research Council, Committee on Nuclear Physics. *Nuclear Physics: The Core of Matter, the Fuel of Stars*. Washington, DC: National Academy Press, 1999.

Olsen, S. L. "Exotic Particles with Four or More Quarks." *Physics Today* (September 2014): 56–57.

Pais, A. "Introducing Atoms and Their Nuclei." In *Twentieth Century Physics*. Vol. 1. Edited by L. M. Brown, A. Pais, and Sir Brian Pippard. Bristol, UK: Institute of Physics Publishing, 1995.

Reinhardt, S., et al. "Test of Relativistic Time Dilation with Fast Optical Atomic Clocks at Different Velocities." *Nature Physics* 3 (2007): 861–64.

Rhodes, R. *The Making of the Atomic Bomb*. New York: Simon and Schuster, 1986.

Rose, P. L. *Heisenberg and the Nazi Atomic Bomb Project: A Study in German Culture*. Berkeley: California University Press, 1998.

Rossi, B. "The Decay of Mesotrons (1939–1943): Experimental Particle Physics in the Age of Innocence." In *The Birth of Particle Physics*. Edited by L. M. Brown and L. Hoddleson, eds. Cambridge: Cambridge University Press, 2010.

Rossi, B., N. Hilberry, and J. Barton Hoag. "The Variation of the Hard Component of Cosmic Rays with Height and the Disintegration of Mesotrons." *Physical Review* 57 (1940): 461–68.

Saarthoff, G., et al. "Improved Test of Time Dilation in Special Relativity." *Physical Review Letters* 91 (2003): 190403-1–190403-4.

Serber, Robert. *The Los Alamos Primer*. Berkeley: University of California Press, 1992.

Sutton, C. *Spaceship Neutrino*. Cambridge: Cambridge University Press, 1992.

Updike, John. *Collected Poems: 1953–1993*. New York: Alfred A. Knopf, 1993. (See p. 313.)

Weisskopf, V. F. "Growing Up with Field Theory: The Development of Quantum Electrodynamics." In *The Birth of Particle Physics*. Edited by L. M. Brown and L. Hoddleson. Cambridge: Cambridge University Press, 2010.

Wikipedia. "Time Dilation." http://en.wikipedia.org/wiki/Time_dilation (accessed April 2, 2014).

Wilcvzek, F. *The Lightness of Being: Mass, Ether and the Unification of Forces*. New York: Basic Books, 2008.

CHAPTER 12: METHODS AND MOTION

Abramowitz, M., and I. A. Stegun. *Handbook of Mathematical Functions with Formulas, Graphs, and Mathematical Tables*. Washington, DC: National Bureau of Standards, 1964.

Akhmediev, N. N. "Déjà Vu in Optics," *Nature* 413 (September 20, 2001): 267–68.

Barger, V. D., and M. G. Olsson. *Classical Mechanics: A Modern Perspective*. 2nd ed. New York: McGraw Hill, 1995. (See ch. 11.)

Berman, G. P. "The Fermi-Pasta-Ulam Problem: Fifty Years of Progress." *CHAOS* 15 (March 2005): 05104-1-18.

Campbell-Kelly, M., M. Croaken, R. Flood, and E. Robson, eds. *The History of Mathematical Tables*. Oxford: Oxford University Press, 2003.

Cannon, J. T., and S. Dostrovsky. *The Evolution of Dynamics: Vibration Theory from 1687 to 1742*. New York: Springer-Verlag, 1981.

Coulson, C. A., and A. Jeffrey. *Waves*. 2nd ed. London: Longman, 1977.

Croaken, M. "Table Making by Committee." In *The History of Mathematical Tables*. Edited by M. Campbell-Kelly, M. Croaken, R. Flood, and E. Robson. Oxford: Oxford University Press, 2003.

————. and M. Campbell-Kelly. "Beautiful Numbers: The Rise and Decline of the British Association Mathematical Tables Committee, 1871–1965." *IEEE Annals of the History of Computing* (October–December 2000): 44–61.

Cvitanovic, P. *Universality in Chaos*. 2nd ed. Bristol, UK: Adam Hilger, 1989.

Dauxois, T. "Fermi, Pasta, Ulam, and a Mysterious Lady." *Physics Today* (January 2008): 55–57.

Dutka, J. "On the Early History of Bessel Functions." *Archive for History of the Exact Sciences* 49 (1995): 105–34.

Fermi, E., J. Pasta, and S. Ulam. "Studies of Nonlinear Problems." *Los Alamos Document LA-1940* (May 1955).

Feynman, R. P. *The Feynman Lectures on Physics*. Reading, MA: Addison-Wesley, 1971. (See chs. 25, 49, and 50 in vol. 1.)

Fricke, W. "Bessel, Friedrich Wilhelm." In *Dictionary of Scientific Biography*. Edited by C. C. Gillespie. New York: Charles Scribner's Sons, 1975.

Gallavotti, G. *The Fermi-Pasta-Ulam Problem: A Status Report*. Berlin: Springer, 2008.

Gleick, J. *Chaos: Making a New Science*. London: Heinemann, 1988.

Grier, D. A. *When Computers Were Human*. Princeton, NJ: Princeton University Press, 2005.

Hall, N., ed. *The New Scientist Guide to Chaos*. London: Penguin, 1991. (Contains a broad range of introductory articles by experts in the field.)

Horsburgh, E. M. *Modern Instruments and Methods of Calculation: A Handbook of the Napier Tercentenary Exhibition*. London: G. Bell and Sons, 1915.

Kevrekidis, P. G. "Non-linear Waves in Lattices: Past, Present, Future." *IMA Journal of Applied Mathematics* 76 (2011): 389–423.

Kibble, T. W. B. *Classical Mechanics*. London: McGraw-Hill, 1966. (See ch. 12.)

Kline, M. *Mathematical Thought from Ancient to Modern Times*. New York: Oxford University Press, 1972.

Magie, W. F. ed. *A Source Book in Physics*. Cambridge, MA: Harvard University Press, 1963.

May, R. M. "Simple Mathematical Models with Very Complicated Dynamics." *Nature* 261 (June 10, 1976): 459–67.

McFarland, D. M. "Quarter Squares Revisited." http//:scholarship.org/uc/item/5n31064n (accessed September 12, 2014).

McLachlan, N. W. *Bessel Functions for Engineers.* 1st ed. Oxford: Clarendon, 1934.

McManus, C. *Right Hand, Left Hand.* London: Phoenix, 2002.

Moon, F. C. *Chaotic Vibrations: An Introduction for Applied Scientists and Engineers.* Hoboken, NJ: John Wiley, 2004.

Motter, A. E., and D. K. Campbell. "Chaos at Fifty." *Physics Today* (May 2013): 27–33.

Norberg, A. L. "Table Making in Astronomy." In *The History of Mathematical Tables.* Edited by M. Campbell-Kelly, M. Croaken, R. Flood, and E. Robson. Oxford: Oxford University Press, 2003.

Pask, C. *Magnificent Principia: Exploring Isaac Newton's Masterpiece.* Amherst, NY: Prometheus Books, 2013. (See ch. 12.)

Pettini, M., L. Casetti, M. Carutti-Sola, R. Franzosi, and E. D. G. Cohen. "Weak and Strong Chaos in Fermi-Pasta-Ulam Models and Beyond." *CHAOS* 15 (March 2005): 05106-1-13.

Porter, M. A., N. J. Zabusky, Bambi Hu, and D. K. Campbell. "Fermi, Pasta, Ulam and the Birth of Experimental Mathematics." *American Scientist* 97 (May–June 2009): 214–21.

Prestini, E. *The Evolution of Applied Harmonic Analysis.* Boston: Birkhäuser, 2004.

Roegel, D. "A Reconstruction of Blater's Table of Quarter-Squares." http://locomat.loria.fr/blater1887/blater1887doc.pdf (accessed September 12, 2014.)

Segre, E., ed. *Collected Papers of Enrico Fermi.* Vol 2. Chicago: University of Chicago Press, 1965.

Stewart, I. *Does God Play Dice? The New Mathematics of Chaos.* London: Penguin Books, 1990.

Struik, D. J., ed. *A Source Book in Mathematics 1200–1800.* Cambridge, MA: Harvard University Press, 1969.

Thomson, Sir William (Lord Kelvin), and P. G. Tait. *Principles of Mechanics and Dynamics* (formerly titled *Treatise on Natural Philosophy*). New York: Dover, 1962. (A republication of the last revised edition of 1912. See sec.75 in vol. 1.)

Wandell, B. A. *Foundations of Vision.* Sunderland, MA: Sinauer, 1995.

Watson, G. N. *A Treatise on the Theory of Bessel Functions.* 2nd ed. Cambridge: Cambridge University Press, 1944.

Westheimer, G. "The Fourier Theory of Vision." *Perception* 30 (2001): 531–41.

CHAPTER 13: EVALUATION

Barron, Rachel Stiffler. *Lise Meitner: Discoverer of Nuclear Fission*. Greensboro, NC: Morgan Reynolds, 2000.

Campbell-Kelly M., M. Croaken, R. Flood, and E. Robson, eds. *The History of Mathematical Tables*. Oxford: Oxford University Press, 2003.

Grier, D. A. "Human Computers: The First Pioneers of the Information Age." *Endeavor* 25 (2001): 28–32.

———. *When Computers Were Human*. Princeton, NJ: Princeton University Press, 2005.

Rife, Patricia. *Lise Meitner and the Dawn of the Nuclear Age*. Boston: Birkhäuser, 1999.

Ronan, C. A. "Edmond Halley." In *Dictionary of Scientific Biography*. Edited by C. C. Gillespie. New York: Charles Scribner's Sons, 1975.

———. *Edmond Halley: Genius in Eclipse*. London: MacDonald, 1970.

Sime, Ruth Lewin. *Lise Meitner: A Life in Physics*. Berkeley: University of California Press, 1997.

Thomson, Sir William (Lord Kelvin), and P. G. Tait. *Principles of Mechanics and Dynamics* (formerly titled *Treatise on Natural Philosophy*). New York: Dover, 1962. (A republication of the last revised edition of 1912. See the appendix.)

Williams, M. R. *A History of Computing Technology*. Englewood Cliffs, NJ: Prentice Hall, 1985.

INDEX OF NAMES

INDEX OF
SUBJECTS AND TERMS